Use R!

Series Editors:
Robert Gentleman   Kurt Hornik   Giovanni Parmigiani

Use R!

Jim Albert

Bayesian Computation with R

 Springer

Professor Jim Albert
Department of Mathematics and Statistics
Bowling Green State University
Bowling Green, OH 43403-0221
USA
albert@bgnet.bgsu.edu

Series Editors:
Robert Gentleman
Program in Computational Biology
Division of Public Health Sciences
Fred Hutchinson Cancer Research Center
1100 Fairview Ave. N, M2-B876
Seattle, Washington, 981029-1024
USA

Kurt Hornik
Department für Statistik und Mathematik
Wirtschaftsuniversität Wien Augasse 2-6
A-1090 Wien
Austria

Giovanni Parmigiani
The Sidney Kimmel Comprehensive
Cancer Center at Johns Hopkins University
550 North Broadway
Baltimore, MD, 21205-2011
USA

Library of Congress Control Number: 2007929182

ISBN: 978-0-387-71384-7   e-ISBN: 978-0-387-71385-4

Printed on acid-free paper.

9 8 7 6 5 4 3 2

springer.com

Preface

There has been dramatic growth in the development and application of Bayesian inference in statistics. Berger (2000) documents the increase in Bayesian activity by the number of published research articles, the number of books, and the extensive number of applications of Bayesian articles in applied disciplines such as science and engineering.

One reason for the dramatic growth in Bayesian modeling is the availability of computational algorithms to compute the range of integrals that are necessary in a Bayesian posterior analysis. Due to the speed of modern computers, it is now possible to use the Bayesian paradigm to fit very complex models that cannot be fit by alternative frequentist methods.

To fit Bayesian models, one needs a statistical computing environment. This environment should be such that one can

- write short scripts to define a Bayesian model
- use or write functions to summarize a posterior distribution
- use functions to simulate from the posterior distribution
- construct graphs to illustrate the posterior inference

An environment that meets these requirements is the R system. R provides a wide range of functions for data manipulation, calculation, and graphical displays. Moreover, it includes a well-developed, simple programming language that users can extend by adding new functions. Many such extensions of the language in the form of packages are easily downloadable from the Comprehensive R Archive Network (CRAN).

The purpose of this book is to introduce Bayesian modeling by the use of computation using the R language. At Bowling Green State University, I have taught an introductory Bayesian inference class to students in masters and doctoral programs in statistics for which this book would be appropriate. This book would serve as a useful companion to the introductory Bayesian texts Gelman et al (2003), Carlin and Louis (2000), Press (2003), Gill (2002), or Lee (2004). Also the book would be valuable to the statistical practitioner who wishes to learn more about the R language and Bayesian methodology.

Chapters 2, 3, and 4 illustrate the use of R for Bayesian inference for standard one- and two-parameter problems. These chapters discuss the use of different types of priors, the use of the posterior distribution to perform different types of inference, and the use of the predictive distribution. The base package of R provides functions to simulate from all of the standard probability distributions, and these functions can be used to simulate from a variety of posterior distributions. Modern Bayesian computing is introduced in Chapters 5 and 6. Chapter 5 discusses the summarization of posterior distribution by posterior modes and introduces rejection sampling and the Monte Carlo approach for computing integrals. Chapter 6 introduces the fundamental ideas of Markov chain Monte Carlo (MCMC) methods and the use of MCMC output analysis to decide if the batch of simulated draws provides a reasonable approximation to the posterior distribution of interest. The remaining chapters illustrate the use of these computational algorithms for a variety of Bayesian applications. Chapter 7 introduces the use of exchangeable models in the simultaneous estimation of a set of Poisson rates. Chapter 8 describes Bayesian tests of simple hypotheses and the use of Bayes factors in comparing models. Chapter 9 describes Bayesian regression models, and Chapter 10 describes several applications, such as robust modeling, binary regression with a probit link, and order-restricted inference that are well-suited for the Gibbs sampling algorithm. Chapter 11 describes the use of R to interface with WinBUGS, a popular program for implementing MCMC algorithms.

An R package, LearnBayes, available from the CRAN site, has been written to accompany this text. This package contains all of the Bayesian R functions and datasets described in the book. One goal in writing LearnBayes is to provide guidance for the student and applied statistician in writing short R functions for implementing Bayesian calculations for their specific problems. Also the LearnBayes package will make it easier for users to use the growing number of R packages for fitting a variety of Bayesian models.

I would like to express my appreciation for the people who provided assistance in preparing this book. John Kimmel, my editor, was most helpful in encouraging me to write this book and provide helpful feedback. I am appreciative of Patricia Williamson and Sherwin Toribio for providing useful suggestions. I am appreciative to all of the students at Bowling Green who have enrolled in my Bayesian statistics class over the years. Finally, but certainly not least, I wish to thank my wife Anne and my children Lynne, Bethany and Steven for encouragement and inspiration.

Bowling Green, Ohio,   *Jim Albert*
                                                                    January 2007

Contents

1

An Introduction to R

1.1 Overview

R is a rich environment for statistical computing and has many capabilities to explore data in its base package. In addition, R contains a collection of functions for simulating and summarizing the familiar one-parameter probability distributions. One goal of this chapter is to provide a brief introduction to basic commands for summarizing and graphing data. We illustrate these commands on a dataset about students in a an introductory statistics class. A second goal of this chapter is to introduce the use of R as an environment for programming Monte Carlo simulation studies. We describe a simple Monte Carlo study to explore the behavior of the two-sample t statistic when testing from populations that deviate from the usual assumptions. We will find these data analysis and simulation commands very helpful in Bayesian computation.

1.2 Exploring a Student Dataset

1.2.1 Introduction to the Dataset

To illustrate some basic commands for summarizing and graphing data, we consider answers from a sheet of questions given to all students in an introductory statistics class at Bowling Green State University. Some of the questions that were asked included:

1. What is your gender?
2. What is your height in inches?
3. Choose a whole number between 1 and 10.
4. Give the time you went to bed last night.
5. Give the time you woke up this morning.
6. What was the cost (in dollars) of your last haircut including the tip?
7. Do you prefer water, pop, or milk with your evening meal?

This is a rich dataset that can be used to illustrate methods for exploring a single batch of categorical or quantitative data, for comparing subgroups of the data, such as comparing the haircut costs of men and women, and for exploring relationships.

1.2.2 Reading the Data into R

The data for 657 students were recorded in a spreadsheet and saved as the file "studentdata.txt" in text format with tabs between the fields. The first line of the data file is a header that includes the variable names.

One can read this data into R by the `read.table` command. There are three arguments used in this command. The first argument is the name of the datafile in quotes; the next argument, `sep`, indicates that fields in the file are separated by tab characters; and the `header=TRUE` argument indicates that the file has a header line with the variable names. This dataset is stored in the R data frame called `studentdata`.

```
> studentdata = read.table("studentdata.txt", sep = "\t",
+    header = TRUE)
```

This dataset is also available as part of the LearnBayes package. Assuming that the package has been installed and loaded into R, one accesses the data by means of the `data` command:

```
> data(studentdata)
```

To see the variable names, we display the first row of the data frame by the `studentdata[1,]` command.

```
> studentdata[1, ]

  Student Height Gender Shoes Number Dvds ToSleep WakeUp
1       1     67 female    10      5   10    -2.5    5.5
  Haircut Job Drink
1      60  30 water
```

To make the variable names visible in the R environment, we use the `attach` command.

```
> attach(studentdata)
```

1.2.3 R Commands to Summarize and Graph a Single Batch

One categorical variable in this dataset is `Drink` which indicates the student's drinking preference between milk, pop, and water. One can tally the different responses by the `table` command.

```
> table(Drink)
```

```
Drink
 milk    pop water
  113    178    355
```

We see that more than half the students preferred water, and pop was more popular than milk.

One can graph these frequencies with a bar graph by the `barplot` command. We first save the output of `table` in the variable `t` and then use `barplot` with `t` as an argument. We add labels to the horizontal and vertical axes by the `xlab` and `ylab` argument options. Fig. 1.1 displays the resulting graph.

```
> t=table(Drink)
> barplot(t,xlab="Drink",ylab="Count")
```

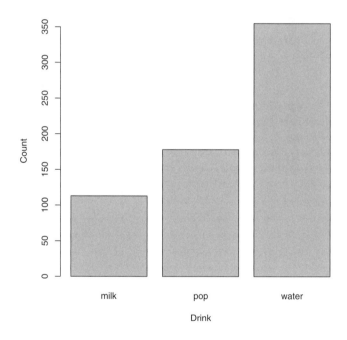

Fig. 1.1. Bar plot of the drinking preference of the statistics students.

Suppose we are next interested in examining how long the students slept the previous night. We did not directly ask the students about their sleeping time, but we can compute a student's hours of sleep by subtracting her go-to-bed time from her wakeup time. In R we perform this computation for all students and the variable `hours.of.sleep` contains the sleeping times.

```
> hours.of.sleep = WakeUp - ToSleep
```

A simple way to summarize this quantitative variable is by the `summary` command that gives a variety of descriptive statistics about the variable.

```
> summary(hours.of.sleep)
```

```
   Min. 1st Qu.  Median    Mean 3rd Qu.    Max.   NA's
  2.500   6.500   7.500   7.385   8.500  12.500  4.000
```

On average, we see that students slept 7.5 hours and half of the students slept between 6.5 and 8.5 hours.

To see the distribution of sleeping times, we can construct a histogram using the `hist` command (see Fig. 1.2).

```
> hist(hours.of.sleep,main="")
```

The shape of this distribution looks symmetric about the average value of 7.5 hours.

1.2.4 R Commands to Compare Batches

Since the gender of each student was recorded, one can make comparisons between men and women on any of the quantitative variables. Do men tend to sleep longer than women? We can answer this question graphically by constructing parallel boxplots of the sleeping times of men and women. Parallel boxplots can be displayed by the `boxplot` command. The argument is given by

```
    hours.of.sleep ~ Gender
```

This indicates that a boxplot of the hours of sleep will be constructed for each level of `Gender`. The resulting graph is displayed in Fig. 1.3. From the display, it appears that men and women are similar with respect to their sleeping time.

```
> boxplot(hours.of.sleep~Gender)
> title(ylab="Hours of Sleep")
```

For other variables, there are substantial differences between the two genders. Suppose we wish to divide the haircut prices into two groups – the haircut prices for the men and the haircut prices for the women. We do this by use of the R logical operator ==. The syntax

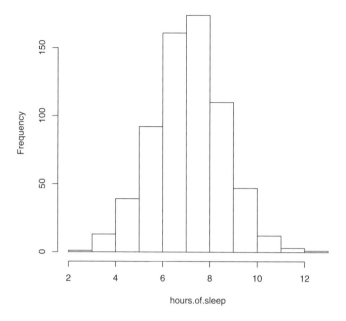

Fig. 1.2. Histogram of the hours of sleep of the statistics students.

    Gender=="female"

is a logical statement that will be TRUE if Gender is "female"; otherwise it will be FALSE. The expression

    Haircut[condition]

will produce a subset of Haircut according to when the condition is TRUE. So the statement

```
> female.Haircut=Haircut[Gender=="female"]
```

will select the haircut prices only for the female students and store the prices into the variable female.Haircut. Similarly, we use the logical operator to store the male haircut prices into the variable male.Haircut.

```
> male.Haircut=Haircut[Gender=="male"]
```

By use of the summary command, we summarize the haircut prices of the women and the men.

```
> summary(female.Haircut)
```

Min.	1st Qu.	Median	Mean	3rd Qu.	Max.	NA's
0.00	15.00	25.00	34.08	45.00	180.00	19.00

The slope is approximately −0.5, which means that a student loses about a half-hour of sleep for every hour later that he or she goes to bed.

We can display this line on top of the scatterplot by the abline command (see Fig. 1.5) . There are two arguments to the function, the intercept and slope of the line to be plotted.

```
> abline(7.9628,-0.5753)
```

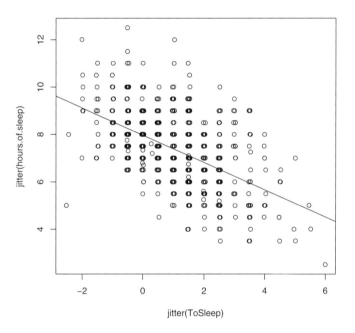

Fig. 1.5. Scatterplot of wake-up time and hours of sleep for students with least-squares line plotted on top.

1.3 Exploring the Robustness of the t Statistic

1.3.1 Introduction

Suppose one has two independent samples $x_1, ..., x_m$ and $y_1, ..., y_n$, and one wishes to test the hypothesis that the mean of the x population is equal to the mean of the y population:

```
        n.reject=n.reject+1 # reject if |t| exceeds critical pt
    }
    true.sig.level=n.reject/N # est. is proportion of rejections
```

1.3.4 The Behavior of the True Significance Level Under Different Assumptions

The R script described in the previous section can be used to explore the pattern of the true significance level α^T for different choices of sample sizes and populations. The only two lines that need to be changed in the R script are the definition of the sample sizes m and n and the two lines where the two samples are simulated.

Suppose we fix the stated significance level at $\alpha = .10$ and keep the sample sizes at $m = 10$ and $n = 10$. We simulate samples from the following populations, where the only restriction is that the population means are equal.

- Normal populations with zero means and equal spreads ($\sigma_x = \sigma_y = 1$).

  ```
  x=rnorm(m,mean=0,sd=1)
  y=rnorm(n,mean=0,sd=1)
  ```

- Normal populations with zero means and very different spreads ($\sigma_x = 1, \sigma_y = 10$).

  ```
  x=rnorm(m,mean=0,sd=1)
  y=rnorm(n,mean=0,sd=10)
  ```

- T populations, 4 degrees of freedom and equal spreads

  ```
  x=rt(m,df=4)
  y=rt(n,df=4)
  ```

- Exponential populations with $\mu_x = \mu_y = 1$.

  ```
  x=rexp(m,rate=1)
  y=rexp(n,rate=1)
  ```

- One normal population ($\mu_x = 10, \sigma_x = 2$) and one exponential population ($\mu_y = 10$).

  ```
  x=rnorm(m,mean=10,sd=2)
  y=rexp(n,rate=1/10)
  ```

The R script was run for each of these five population scenarios using $N = 10,000$ iterations and the estimated true significance levels are displayed in Table 1.1. These values should be compared with the stated significance level of $\alpha = .1$, keeping in mind that the simulation standard error of each estimate is equal to .003. (The simulation standard error, the usual standard error in computing a binomial proportion, is equal to $\sqrt{.1(.9)/10000} = 0.003$.)

1.3.3 Programming a Monte Carlo Simulation

Suppose we are interested in learning about the true significance level for the t statistic when the populations don't follow the standard assumptions of normality and equal variances. In general, the true significance level will depend on

- the stated level of significance α
- the shape of the populations (normal, skewed, heavy-tailed, etc.)
- the spreads of the two populations as measured by the two standard deviations
- the sample sizes m and n

Given a particular choice of α, shape, spreads, and sample sizes, we wish to estimate the true significance level given by

$$\alpha^T = P(|T| \geq t_{n+m-2,\alpha/2}).$$

Here is an outline of a simulation algorithm to compute α^T:

1. Simulate a random sample $x_1, ..., x_m$ from the first population and $y_1, ..., y_n$ from the second population.
2. Compute the t statistic T from the two samples.
3. Decide if $|T|$ exceeds the critical point and H_0 is rejected.

One repeats steps 1–3 of the algorithm N times. One estimates the true significance level by

$$\hat{\alpha}^T = \frac{\text{number of rejections of } H_0}{N}.$$

The following is an R script that implements the simulation algorithm for normal populations with mean 0 and standard deviation 1. The R variable alpha is the stated significance level, m and n are the sample sizes, and N is the number of simulations. The rnorm command is used to simulate the two samples and t contains the value of the t statistic. One decides to reject if

```
abs(t)>qt(1-alpha/2,n+m-2)
```

where qt(p,df) is the pth quantile of a t distribution with df degrees of freedom. The observed significance level is stored in the variable true.sig.level.

```
alpha=.1; m=10; n=10          # sets alpha, m, n
N=10000                        # sets the number of simulations
n.reject=0                     # counter of num. of rejections
for (i in 1:N)
{
    x=rnorm(m,mean=0,sd=1)     # simulates xs from population 1
    y=rnorm(n,mean=0,sd=1)     # simulates ys from population 2
    t=tstatistic(x,y)          # computes the t statistic
    if (abs(t)>qt(1-alpha/2,n+m-2))
```

50 and standard deviation 10 using the **rnorm** function and store the vector of values in the variable x. Likewise we simulate a sample of ys by simulating 10 values from a $N(50, 10)$ distribution and store these values in the variable y.

```
> x=rnorm(10,mean=50,sd=10)
> y=rnorm(10,mean=50,sd=10)
```

Next we write a few lines of R code to compute the value of the t statistic from the samples in x and y. We find the sample sizes m and n by the R **length** command.

```
> m=length(x)
> n=length(y)
```

We compute the pooled standard deviation s_p – in the R code, **sd** is the standard deviation function and **sqrt** takes the square root of its argument.

```
> sp=sqrt(((m-1)*sd(x)^2+(n-1)*sd(y)^2)/(m+n-2))
```

With m, n, and sp defined, we compute the t statistic

```
> t=(mean(x)-mean(y))/(sp*sqrt(1/m+1/n))
```

By combining these R statements, we can write a short R function **tstatistic** to compute the t statistic. This function has two arguments, the vectors x and y, and the output of the function (indicated by the **return** statement) is the value of the t statistic.

```
tstatistic=function(x,y)
{
m=length(x)
n=length(y)
sp=sqrt(((m-1)*sd(x)^2+(n-1)*sd(y)^2)/(m+n-2))
t=(mean(x)-mean(y))/(sp*sqrt(1/m+1/n))
return(t)
}
```

Suppose this function has been saved in the file "tstatistic.R". We enter this function into R by means of the **source** command.

```
> source("tstatistic.R")
```

We try the function by placing some fake data in vectors **data.x** and **data.y** and then computing the t statistic on these data:

```
> data.x=c(1,4,3,6,5)
> data.y=c(5,4,7,6,10)
> tstatistic(data.x, data.y)

[1] -1.937926
```

$$H_0 : \mu_x = \mu_y.$$

Let $\bar{X}$ and $\bar{Y}$ denote the sample means of the xs and ys and let s_x and s_y denote the respective standard deviations. The standard test of this hypothesis H_0 is based on the t statistic

$$T = \frac{\bar{X} - \bar{Y}}{s_p\sqrt{1/m + 1/n}},$$

where s_p is the pooled standard deviation

$$s_p = \sqrt{\frac{(m-1)s_x^2 + (n-1)s_y^2}{m+n-2}}.$$

Under the hypothesis H_0, the test statistic T has a t distribution with $m+n-2$ degrees of freedom when

- both the xs and ys are independent random samples from normal distributions
- the standard deviations of the x and y populations, σ_x and σ_y, are equal

Suppose the level of significance of the test is set at α. Then one will reject H when

$$|T| \geq t_{n+m-2,\alpha/2},$$

where $t_{df,\alpha}$ is the $(1-\alpha)$ quantile of a t random variable with df degrees of freedom.

If the underlying assumptions of normal populations and equal variances hold, then the level of significance of the t-test will be the stated level of α. But in practice, many people use the t statistic to compare two samples even when the underlying assumptions are in doubt. So an interesting problem is to investigate the robustness or sensitivity of this popular test statistic with respect to changes in the assumptions. If the stated significance level is $\alpha = .10$ and the populations are skewed or have heavy tails, what will be the true significance level? If the assumption of equal variances is violated and there are significant differences in the spreads of the two populations, what is the true significance level? One can answer these questions through a Monte Carlo simulation study. R is a very suitable platform for writing a simulation algorithm. One can generate random samples from a wide variety of probability distributions and R has an extensive set of data analysis capabilities for summarizing and graphing the simulation output. Here we illustrate the construction of a simple R function to address the robustness of the t statistic.

1.3.2 Writing a Function to Compute the t Statistic

To begin, we generate some random data for the samples of xs and ys. We simulate a sample of 10 observations from a normal distribution with mean

In this brief study, it appears that if the populations have equal spreads, then the true significance level is approximately equal to the stated level for different population shapes. If the populations have similar shapes and different spreads, then the true significance level can be slightly higher than 10%. If the populations have substantially different shapes (such as normal and exponential) and unequal spreads, then the true significance level can be substantially higher than the stated level.

Table 1.1. True significance levels of the t-test computed by Monte Carlo experiments. The standard error of each estimate is approximately 0.003.

Populations	True Significance Level
Normal populations with equal spreads	0.0986
Normal populations with unequal spreads	0.1127
t(4) distributions with equal spreads	0.0968
Exponential populations with equal spreads	0.1019
Normal and exponential populations with unequal spreads	0.1563

Since the true significance level in the last case is 50% higher than the stated level, one might be interested in seeing the exact sampling distribution of the t statistic. We rerun this simulation for the normal and exponential populations, storing the simulated values of the t statistic in a vector `tstat`. We use the R command `density` to construct a nonparametric density estimate of the exact sampling distribution of the t statistic. The `lines` command is used to plot the t density with 18 degrees of freedom on top. Fig. 1.6 displays the resulting graph of the two densities. Note that the actual sampling distribution of the t statistic is right-skewed, which would account for the large true significance level.

```
> plot(density(tstat),xlim=c(-5,8),ylim=c(0,.4),lwd=3)
> lines(x,dt(x,df=18))
> legend(4,.3,c("exact","t(18)"),lwd=c(3,1))
```

1.4 Further Reading

Although R is a sophisticated package with many commands, there are many resources available for learning the package. There is some basic instruction on R that can be found from the R Help menu. The R project home page at http://www.r-project.org lists a number of books describing different levels of statistical computing using R. Verzani (2004) is a good book describing the use of R in an introductory statistics course; in particular, the book is helpful

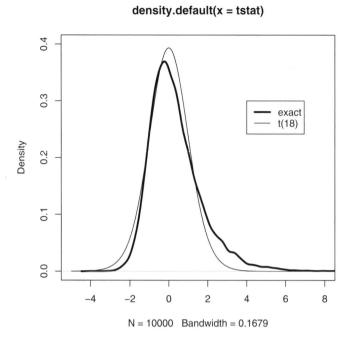

Fig. 1.6. Exact sampling density of the t statistic when sampling from normal and exponential distributions.

for getting started in constructing different types of graphical displays. Gentle (2002), Appendix A, gives a general description of Monte Carlo experiments with an extended example.

1.5 Summary of R Functions

An outline of the R functions used in this chapter is presented here. Detailed information about any specific function, say `abline`, can be found by typing

`?abline`

in the R command window.

`abline` – add a straight line to a plot

`attach` – attach a set of R objects to search path

`barplot` – create a barplot with vertical or horizontal bars

`boxplot` – produce box-and-whisker plot(s) of the given (grouped) values

`density` – computes kernel density estimates

`hist` – computes a histogram of the given data values

`lm` – used to fit linear models such as regression

`mean` – computes the arithmetic mean

`plot` – generic function for plotting R objects

`read.table` – reads a file in table format and creates a data frame from it, with cases corresponding to lines and variables to fields in the file

`rexp` – random generation for the exponential distribution

`rnorm` – random generation for the normal distribution

`rt` – random generation for the t distribution

`sd` – computes the value of the standard deviation

`summary` – generic function used to produce result summaries of the results of various model fitting functions

`table` – uses the cross-classifying factors to build a contingency table of the counts at each combination of factor levels

1.6 Exercises

1. **Movie DVDs owned by students**
   The variable `Dvds` in the student dataset contains the number of movie DVDs owned by students in the class.
   a) Construct a histogram of this variable by use of the `hist` command.
   b) Summarize this variable by the `summary` command.
   c) Use the `table` command to construct a frequency table of the individual values of `Dvds` that were observed. If one constructs a barplot of these tabled values by use of the command
   ```
   barplot(table(Dvds))
   ```
   one will see that particular response values are very popular. Is there any explanation for these popular values for number of DVDs owned?
2. **Student heights**
   The variable `Height` contains the height (in inches) of each student in the class.
   a) Construct parallel boxplots of the heights by the `Gender` variable.
   b) If one assigns the boxplot output to a variable
   ```
   output=boxplot(Height~Gender)
   ```
   then `output` is a list that contains statistics used in constructing the boxplots. Print `output` to see the statistics that are stored.
   c) On average, how much taller are male students than female students?

3. **Sleeping times**

   The variables `ToSleep` and `WakeUp` contain, respectively, the time to bed and wakeup time for each student the previous evening. (The data are recorded as hours past midnight, so a value of -2 indicates 10 p.m.)

   a) Construct a scatterplot of `ToSleep` and `WakeUp`.
   b) Find a least-squares fit to these data by the `lm` command.
   c) Place the least-squares fit on the scatterplot by the `abline` command.
   d) Use the line to predict the wakeup time for a student who went to bed at midnight.

4. **Performance of the traditional confidence interval for a proportion**

   Suppose one observes y that is binomially distributed with sample size n and probability of success p. The standard 90% confidence interval for p is given by

   $$C(y) = (\hat{p} - 1.645\sqrt{\frac{\hat{p}(1 - \hat{p})}{n}}, \hat{p} + 1.645\sqrt{\frac{\hat{p}(1 - \hat{p})}{n}}),$$

   where $\hat{p} = y/n$. We use this procedure under the assumption that

   $$P(p \in C(y)) = 0.90 \text{ for all } 0 < p < 1.$$

   The function `binomial.conf.interval` will return the limits of a 90% confidence interval given values of y and n.

   ```
   binomial.conf.interval=function(y,n)
   {
   z=qnorm(.95)
   phat=y/n
   se=sqrt(phat*(1-phat)/n)
   return(c(phat-z*se,phat+z*se))
   }
   ```

   a) Read the function `binomial.conf.interval` into R.
   b) Suppose that samples of size $n = 20$ are taken and the true value of the proportion is $p = .5$. Using the `rbinom` command, simulate a value of y and use `binomial.conf.interval` to compute the 90% confidence interval. Repeat this a total of 20 times and estimate the true probability of coverage $P(p \in C(y))$.
   c) Suppose that $n = 20$ and the true value of the proportion is $p = .05$. Simulate 20 binomial random variates with $n = 20$ and $p = .05$ and for each simulated y, compute a 90% confidence interval. Estimate the true probability of coverage.

5. **Performance of the traditional confidence interval for a proportion**

   Exercise 4 demonstrated that the actual probability of coverage of the traditional confidence interval depends on the values of n and p. Construct a Monte Carlo study that investigates how the probability of coverage depends on the sample size and true proportion value. In the study, let n be 10, 25, and 100, and let p be .05, .25, and .50. Write an R function that has three inputs: n, p, and the number of Monte Carlo simulations m, and will output the estimate of the exact coverage probability. Implement your function using each combination of n and p and $m = 1000$ simulations. Describe how the actual probability of converge of the traditional interval depends on the sample size and true proportion value.

The value of the proportion p is unknown. In the Bayesian viewpoint a person's beliefs about the uncertainty in this proportion are represented by a probability distribution placed on this parameter. This distribution reflects the person's subjective prior opinion about plausible values of p.

A random sample of students from a particular university will be taken to learn about this proportion. But first the person does some initial research to learn about the sleeping habits of college students. This research will help her in constructing a prior distribution.

In the Internet article "College Students Don't Get Enough Sleep" in *The Gamecock*, the student newspaper of the University of South Carolina (April 20, 2004), the person reads that a sample survey reports that most students spend only six hours sleeping. She reads a second article "Sleep on It: Implementing a Relaxation Program into the College Curriculum" in *Fresh Writing*, a 2003 publication of the University of Notre Dame. Based on a sample of 100 students, "approximately 70% reported receiving only five to six hours of sleep on the weekdays, 28% receiving seven to eight, and only 2% receiving the healthy nine hours for teenagers."

Based on this information, this person believes that college students generally get less than eight hours of sleep and so p (the proportion that sleep at least eight hours) is likely smaller than .5. After some reflection, her best guess at the value of p is .3. But it is very plausible that this proportion could be any value in the interval from 0 to .5.

A sample of 27 students is taken – in this group, 11 record that they had at least eight hours of sleep the previous night. Based on the prior information and this observed data, the person is interested in estimating the proportion p. In addition, she is interested in predicting the number of students that get at least eight hours of sleep if a new sample of 20 students is taken.

Suppose that our prior density for p is denoted by $g(p)$. If we regard a "success" as sleeping at least eight hours and we take a random sample with s successes and f failures, then the likelihood function is given by

$$L(p) = p^s(1-p)^f, \ 0 < p < 1.$$

The posterior density for p, by Bayes' rule, is obtained, up to a proportionality constant, by multiplying the prior density by the likelihood.

$$g(p|\text{data}) \propto g(p)L(p).$$

We demonstrate posterior distribution calculations using three different choices of the prior density g corresponding to three methods for representing the person's prior knowledge about the proportion.

2.3 Using a Discrete Prior

A simple approach for assessing a prior for p is to write down a list of plausible proportion values and then assign weights to these values. In our example, the person believes that

$$.05, .15, .25, .35, .45, .55, .65, .75, .85, .95$$

are possible values for p. Based on her beliefs, she assigns these values the corresponding weights

$$2, 4, 8, 8, 4, 2, 1, 1, 1, 1,$$

which can be converted to prior probabilities by dividing each weight by the sum. In R, we define p to be the vector of proportion values and prior the corresponding weights that we normalize to probabilities. The plot command is used with the "histogram" type option to graph the prior distribution, and Fig. 2.1 displays the graph.

```
> p = seq(0.05, 0.95, by = 0.1)
> prior = c(2, 4, 8, 8, 4, 2, 1, 1, 1, 1)
> prior = prior/sum(prior)
> plot(p, prior, type = "h", ylab="Prior Probability")
```

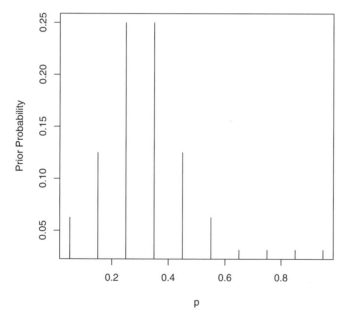

Fig. 2.1. A discrete prior distribution for a proportion p.

In our example, 11 of 27 students sleep a sufficient number of hours, so $s = 11$ and $f = 16$, and the likelihood function is

$$L(p) = p^{11}(1 - p)^{16}, \ 0 < p < 1.$$

(Note that the likelihood is a beta density with parameters $s + 1 = 12$ and $f + 1 = 17$.) The R function pdisc in the package LearnBayes computes the posterior probabilities. To use pdisc, one inputs the vector of proportion values p, the vector of prior probabilities prior, and a data vector data consisting of s and f. The output of pdisc is a vector of posterior probabilities. The cbind command is used to display a table of the prior and posterior probabilities, and Fig. 2.2 displays a line graph of the posterior probabilities.

```
> data = c(11, 16)
> post = pdisc(p, prior, data)
> cbind(p, prior, post)

          p    prior           post
 [1,]  0.05  0.06250  2.882642e-08
 [2,]  0.15  0.12500  1.722978e-03
 [3,]  0.25  0.25000  1.282104e-01
 [4,]  0.35  0.25000  5.259751e-01
 [5,]  0.45  0.12500  2.882131e-01
 [6,]  0.55  0.06250  5.283635e-02
 [7,]  0.65  0.03125  2.976107e-03
 [8,]  0.75  0.03125  6.595185e-05
 [9,]  0.85  0.03125  7.371932e-08
[10,]  0.95  0.03125  5.820934e-15

> plot(p, post, type = "h", ylab="Posterior Probability")
```

Here we note that most of the posterior probability is concentrated on the values $p = .35$ and $p = .45$. If we combine the probabilities for the three most likely values, we can say the posterior probability that p falls in the set $\{.25, .35, .45\}$ is equal to .942.

2.4 Using a Beta Prior

Since the proportion is a continuous parameter, an alternative approach is to construct a density $g(p)$ on the interval $(0, 1)$ that represents the person's initial beliefs. Suppose she believes that the proportion is equally likely to be smaller or larger than $p = .3$. Moreover, she is 90% confident that p is less than .5. A convenient family of densities for a proportion is the beta with kernel proportional to

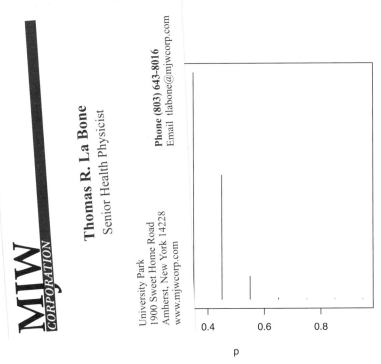

Fig. 2.2. Posterior distribution for a proportion p using a discrete prior.

$$g(p) \propto p^{a-1}(1-p)^{b-1}, \ 0 < p < 1,$$

where the hyperparameters a and b are chosen to reflect the user's prior beliefs about p. Here the person believes that the median and 90th percentiles are given, respectively, by .3 and .5, and this can be matched, by trial and error, with a beta density with $a = 3.4$ and $b = 7.4$. Combining this beta prior with the likelihood function, one can show that the posterior density is also of the beta form with updated parameters $a + s$ and $b + f$.

$$g(p|\text{data}) \propto p^{a+s-1}(1-p)^{b+f-1}, \ 0 < p < 1,$$

where $a + s = 3.4 + 11$ and $b + f = 7.4 + 16$. (This is an example of a conjugate analysis where the prior and posterior densities have the same functional form.) Since the prior, likelihood, and posterior are all in the beta family, we can use the R command `dbeta` to compute values of prior, likelihood, and posterior. These three densities are displayed using the R commands `plot` and `lines` in the same graph in Fig. 2.3. This figure is helpful in seeing that the posterior density in this case compromises between the initial prior beliefs and the information in the data.

```
> p = seq(0, 1, length = 500)
> a = 3.4
> b = 7.4
> s = 11
> f = 16
> prior=dbeta(p,a,b)
> like=dbeta(p,s+1,f+1)
> post=dbeta(p,a+s,b+f)
> plot(p,post,type="l",ylab="Density",lty=2,lwd=3)
> lines(p,like,lty=1,lwd=3)
> lines(p,prior,lty=3,lwd=3)
> legend(.7,4,c("Prior","Likelihood","Posterior"),
+      lty=c(3,1,2),lwd=c(3,3,3))
```

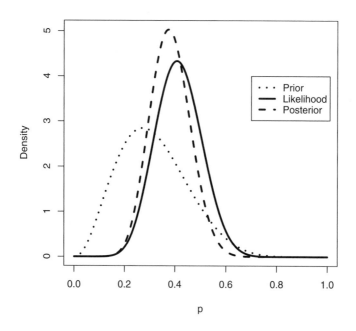

Fig. 2.3. The prior density $g(p)$, the likelihood function $L(p)$, and the posterior density $g(p|\text{data})$ for learning about a proportion p.

We illustrate different ways of summarizing the beta posterior distribution to make inferences about the proportion of heavy sleepers p. The beta cdf and inverse cdf functions `pbeta` and `qbeta` are useful in computing probabilities and constructing interval estimates for p. Is it likely that the proportion of heavy sleepers is greater than .5? This is answered by computing the posterior probability $P(p >= .5|\text{data})$, which is given by the R command

```
> 1 - pbeta(0.5, a + s, b + f)
```

```
[1] 0.0684257
```

This probability is small, so it is unlikely that more than half of the students are heavy sleepers. A 90% interval estimate for p is found by computing the 5th and 95th percentiles of the beta density:

```
> qbeta(c(0.05, 0.95), a + s, b + f)
```

```
[1] 0.2562364 0.5129274
```

We are 90% confident that the proportion is between .256 and .513.

These summaries are exact because they are based on R functions for the beta posterior density. An alternative method of summarization of a posterior density is based on simulation. In this case we can simulate a large number of values from the beta posterior density and summarize the simulated output. Using the random beta command `rbeta`, we simulate 1000 random proportion values from the beta$(a + s, b + f)$ posterior by the command

```
> ps = rbeta(1000, a + s, b + f)
```

and display the posterior as a histogram of the simulated values in Fig. 2.4.

```
> hist(ps,xlab="p",main="")
```

The probability that the proportion is larger than .5 is estimated by the proportion of simulated values in this range.

```
> sum(ps >= 0.5)/1000
```

```
[1] 0.075
```

A 90 percent interval estimate can be estimated by the 5th and 95th sample quantiles of the simulated sample.

```
> quantile(ps, c(0.05, 0.95))
```

```
      5%        95%
0.2599039 0.5172406
```

Note that these summaries of the posterior density for p based on simulation are approximately equal to the exact values based on calculations from the beta distribution.

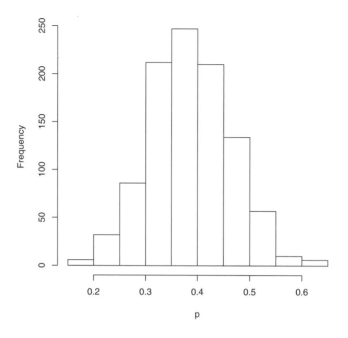

Fig. 2.4. A simulated sample from the beta posterior distribution of p.

2.5 Using a Histogram Prior

Although there are computational advantages to the use of a beta prior, it is straightforward to perform posterior computations for any choice of prior. We outline a "brute-force" method of summarizing posterior computations for an arbitrary prior density $g(p)$.

- Choose a grid of values of p over an interval that covers the posterior density.
- Compute the product of the likelihood $L(p)$ and the prior $g(p)$ on the grid.
- Normalize by dividing each product by the sum of the products. In this step, we are approximating the posterior density by a discrete probability distribution on the grid.
- By use of the R command `sample`, take a random sample with replacement from the discrete distribution.

The resulting simulated draws are an approximate sample from the posterior distribution.

We illustrate this "brute-force" algorithm for a "histogram" prior that may better reflect the person's prior opinion about the proportion p. Suppose it is convenient for our person to state her prior beliefs about the proportion of heavy sleepers by dividing the range of p into 10 subintervals $(0, .1)$, $(.1, .2)$, ... $(.9, 1)$, and then assigning probabilities to the intervals. In our example, the person assigns the weights 2, 4, 8, 8, 4, 2, 1, 1, 1, 1 to these intervals – this can be viewed as a continuous version of the discrete prior used earlier.

In R, we represent this histogram prior with the vector `midpt` that contains the midpoints of the intervals and the vector `prior` that contains the associated prior weights. We convert the prior weights to probabilities by dividing each weight by the sum. We graph this prior in Fig. 2.5 on a grid of values `p` using the R function `histprior` in the LearnBayes package.

```
> midpt = seq(0.05, 0.95, by = 0.1)
> prior = c(2, 4, 8, 8, 4, 2, 1, 1, 1, 1)
> prior = prior/sum(prior)
> p = seq(0, 1, length = 500)

> plot(p,histprior(p,midpt,prior),type="l",
+   ylab="Prior density",ylim=c(0,.25))
```

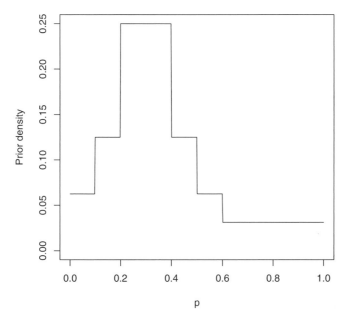

Fig. 2.5. A histogram prior for a proportion p.

On the grid of values of p, we compute the posterior density by multiplying the histogram prior by the likelihood function. (Recall that the likelihood function for binomial density is given by a beta$(s + 1, f + 1)$ density; this function is available by the `dbeta` function.) In Fig. 2.6, the posterior density is displayed.

```
> like = dbeta(p, s + 1, f + 1)
> post = like * histprior(p, midpt, prior)

> plot(p, post, type = "l",ylab="Posterior density")
```

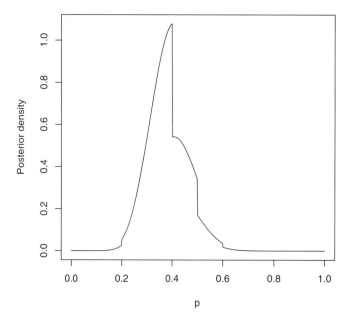

Fig. 2.6. The posterior density for a proportion using a histogram prior

To obtain a simulated sample from the posterior density by our algorithm, we convert the products on the grid to probabilities

```
> post = post/sum(post)
```

and take a sample with replacement from the grid using the R function `sample`.

```
> ps = sample(p, replace = TRUE, prob = post)
```

Fig. 2.7 shows a histogram of the simulated values.

```
> hist(ps, xlab="p")
```

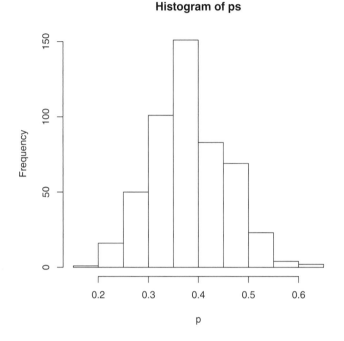

Fig. 2.7. A histogram of simulated draws from the posterior distribution of p with the use of a histogram prior.

The simulated draws can be used as before to summarize any feature of the posterior distribution of interest.

2.6 Prediction

We have focused on learning about the population proportion of heavy sleepers p. Suppose our person is also interested in predicting the number of heavy sleepers $\tilde{y}$ in a future sample of $m = 20$ students. If the current beliefs about p are contained in the density $g(p)$, then the predictive density of $\tilde{y}$ is given by

$$f(\tilde{y}) = \int f(\tilde{y}|p)g(p)dp.$$

We save the frequencies of $\tilde{y}$ in a vector `freq`. Then we convert the frequencies to probabilities by dividing each frequency by the sum and use the `plot` command to graph the predictive distribution (see Fig. 2.8).

```
> freq=table(y)
> ys=c(0:max(y))
> predprob=freq/sum(freq)
> plot(ys,predprob,type="h",xlab="y",
+    ylab="Predictive Probability")
```

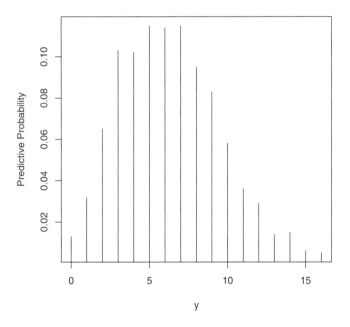

Fig. 2.8. A graph of the predictive probabilities of the number of sleepers $\tilde{y}$ in a future sample of size 20 when the proportion is assigned a beta(3.4, 7.4) prior.

Suppose we wish to summarize this discrete predictive distribution by an interval that covers at least 90% of the probability. The R function `discint` in the LearnBayes package is useful for this purpose. In the output the vector `ys` contains the values of $\tilde{y}$ and `predprob` contains the associated probabilities found from the table output. The matrix `dist` contains the probability distribution with the columns `ys` and `predprob`. The function `discint` has two inputs: the matrix `dist` and a given coverage probability `covprob`. The output is a list where the component `set` gives the confidence set and `prob` gives the exact coverage probability.

```
> dist=cbind(ys,predprob)
```

```
> dist
         ys predprob
 [1,]   0    0.013
 [2,]   1    0.032
 [3,]   2    0.065
 [4,]   3    0.103
 [5,]   4    0.102
 [6,]   5    0.115
 [7,]   6    0.114
 [8,]   7    0.115
 [9,]   8    0.095
[10,]   9    0.083
[11,]  10    0.058
[12,]  11    0.036
[13,]  12    0.029
[14,]  13    0.014
[15,]  14    0.015
[16,]  15    0.006
[17,]  16    0.005
```

```
> covprob=.9
> discint(dist,covprob)
```

```
$prob
[1] 0.918
```

```
$set
[1]  1  2  3  4  5  6  7  8  9 10 11
```

We see that the probability that $\tilde{y}$ falls in the interval $\{1, 2, 3, 4, 5, 6, 7, 8, 9, 10, 11\}$ is 91.8%. To say it in a different way, let $\tilde{y}/20$ denote the proportion of sleepers in the future sample. The probability this sample proportion falls in the interval $[1/20, 11/20]$ is 91.8%. As expected, this interval is much wider than a 91.8% probability interval for the population proportion p. In predicting a future sample proportion, there are two sources of uncertainty, the uncertainty in the value of p and the binomial uncertainty in the value of $\tilde{y}$ and the predictive interval is relatively long since it incorporates both types of uncertainty.

2.7 Further Reading

A number of books are available that describe the basic tenets of Bayesian thinking. Berry (1996) and Albert and Rossman (2001) describe the Bayesian approach for proportions at an introductory statistics level. Albert (1996) describes Bayesian computational algorithms for proportions using the statistics package Minitab. Antleman (1996) and Bolstad (2004) provide elementary descriptions of Bayesian thinking suitable for undergraduate statistics classes.

2.8 Summary of R Functions

discint – computes a highest probability interval for a discrete distribution
Usage: discint(dist,prob)
Arguments: dist, a probability distribution written as a matrix where the first column contains the values and the second column contains the probabilities; prob, the probability content of interest
Value: prob, the exact probability content of the interval, and set, the set of values of the probability interval

histprior – computes the density of a probability distribution defined on a set of equal-width intervals
Usage: histprior(p,midpts,prob)
Arguments: p, the vector of values for which the density is to computed; midpts, the vector of midpoints of the intervals; prob, the vector of probabilities of the intervals
Value: vector of values of the probability density

pdisc – computes the posterior distribution for a proportion for a discrete prior distribution
Usage: pdisc(p, prior, data)
Arguments: p, a vector of proportion values; prior, a vector of prior probabilities; data, a vector consisting of the number of successes and number of failures
Value: the vector of posterior probabilities

pdiscp – computes predictive distribution for the number of successes of a future binomial experiment with a discrete distribution for the proportion
Usage: pdiscp(p, probs, n, s)
Arguments: p, the vector of proportion values; probs, the vector of probabilities; n, the size of the future binomial sample; s, the vector of number of successes for future binomial experiment
Value: the vector of predictive probabilities for the values in the vector s

pbetap – computes predictive distribution for the number of successes of a future binomial experiment with a beta distribution for the proportion
Usage: pbetap(ab, n, s)

Arguments: ab, the vector of parameters of the beta prior; n, the size of the future binomial sample; s, the vector of number of successes for future binomial experiment
Value: the vector of predictive probabilities for the values in the vector s

2.9 Exercises

1. **Estimating a proportion with a discrete prior**
   Bob claims to have ESP. To test this claim, you propose the following experiment. You will select one from four large cards with different geometric figures and Bob will try to identify it. Let p denote the probability that Bob is correct in identifying the figure for a single card. You believe that Bob has no ESP ability ($p = .25$), but there is a small chance that p is either larger or smaller than .25. After some thought, you place the following prior distribution on p:

p	0	.125	.250	.375	.500	.625	.750	.875	1
$g(p)$	.001	.001	.950	.008	.008	.008	.008	.008	.008

   Suppose that the experiment is repeated ten times and Bob is correct six times and incorrect four times. Using the function pdisc, find the posterior probabilities of these values of p. What is your posterior probability that Bob has no ability?

2. **Estimating a proportion with a histogram prior**
   Consider the following experiment. Hold a penny on edge on a flat hard surface, and spin it with your fingers. Let p denote the probability that it lands heads. To estimate this probability, we will use a histogram to model our prior beliefs about p. Divide the interval $[0,1]$ into the 10 subintervals $[0,.1], [.1,.2], ..., [.9,1]$, and specify probabilities that p is in each interval. Next spin the penny 20 times and count the number of successes (heads) and failures (tails). Simulate from the posterior distribution by (1) computing the posterior density of p on a grid of values on $(0, 1)$ and (2) taking a simulated sample with replacement from the grid. (The functions histprior and sample are helpful in this computation.) How have the interval probabilities changed on the basis of your data?

3. **Estimating a proportion and prediction of a future sample**
   A study reported on the long-term effects of exposure to low levels of lead in childhood. Researchers analyzed children's shed primary teeth for lead content. Of the children whose teeth had a lead content of more than 22.22 parts per million (ppm), 22 eventually graduated from high school and 7 did not. Suppose your prior density for p, the proportion of all such children who will graduate from high school is beta(1, 1), and so your posterior density is beta(23, 8).
   a) Use the function qbeta to find a 90% interval estimate for p.
   b) Use the function pbeta to find the probability that p exceeds .6.

c) Use the function rbeta to take a simulated sample of size 1000 from the posterior distribution of p.

d) Suppose you find 10 more children who have a lead content of more than 22.22 ppm. Find the predictive probability that 9 or 10 of them will graduate from high school. (Use your simulated sample from part (c) and the rbinom function to take a simulated sample from the predictive distribution.)

4. **Contrasting predictions using two different priors**
Suppose two persons are interested in estimating the proportion p of students at a college who commute to school. Suppose Joe uses a discrete prior given in the following table:

p	0.1	0.2	0.3	0.4	0.5
$g(p)$	0.5	0.2	0.2	0.05	0.05

Sam decides instead to use a beta(3, 12) prior for the proportion p.

a) Use R to compute the mean and standard deviation of p for Joe's prior and for Sam's prior. Based on this computation, do Joe and Sam have similar prior beliefs about the location of p?

b) Suppose one is interested in predicting the number of commuters y in a future sample of size 12. Use the functions pdiscp and pbetap to compute the predictive probabilities of y using both Joe's prior and Sam's prior. Do the two people have similar beliefs about the outcomes of a future sample?

5. **Estimating a normal mean with a discrete prior**
Suppose you arc interested in estimating the average total snowfall per year μ (in inches) for a large city on the East Coast of the United States. Assume individual yearly snow totals $y_1, ..., y_n$ are collected from a population that is assumed to be normally distributed with mean μ and known standard deviation $\sigma = 10$ inches.

a) Before collecting data, suppose you believe that the mean snowfall μ can be the values 20, 30, 40, 50, 60, and 70 inches with the following probabilities:

μ	20	30	40	50	60	70
$g(\mu)$	.1	.15	.25	.25	.15	.1

Place the values of μ in the vector mu and the associated prior probabilities in the vector prior.

b) Suppose you observe the yearly snowfall totals 38.6, 42.4, 57.5, 40.5, 51.7, 67.1, 33.4, 60.9, 64.1, 40.1, 40.7, and 6.4. Enter these data into a vector y and compute the sample mean ybar.

c) In this problem, the likelihood function is given by

$$L(\mu) = \exp(-\frac{n}{2\sigma^2}(\mu - \bar{y})^2),$$

where $\bar{y}$ is the sample mean. Compute the likelihood on the list of values in mu and place the likelihood values in the vector like.

d) One can compute the posterior probabilities for μ using the formula
```
post=prior*like/sum(prior*like)
```
Compute the posterior probabilities of μ for this example.

e) Using the function `discint`, find an 80% probability interval for μ.

6. **Estimating a Poisson mean using a discrete prior** (from Antleman (1996))

Suppose you own a trucking company with a large fleet of trucks. Break-downs occur randomly in time and the number of breakdowns during an interval of t days is assumed to be Poisson distributed with mean $t\lambda$. The parameter λ is the daily breakdown rate. The possible values for λ are .5, 1, 1.5, 2, 2.5, and 3 with respective probabilities .1, .2, .3, .2, .15, and .05. If one observes y breakdowns, then the posterior probability of λ is proportional to

$$g(\lambda)\exp(-t\lambda)(t\lambda)^y,$$

where g is the prior probability.

a) If 12 trucks break down in a six-day period, find the posterior probabilities for the different rate values.

b) Find the probability that there are no breakdowns during the next week. Hint: If the rate is λ, the conditional probability of no break-downs during a seven-day period is given by $\exp\{-7\lambda\}$. One can compute this predictive probability by multiplying a list of conditional probabilities by the posterior probabilities of λ and finding the sum of the products.

3
Single-Parameter Models

3.1 Introduction

In this chapter, we introduce the use of R in summarizing the posterior distributions for several single-parameter models. We begin by describing Bayesian inference for a variance for a normal population and inference for a Poisson mean when informative prior information is available. For both problems, summarization of the posterior distribution is facilitated by the use of R functions to compute and simulate distributions from the exponential family. In Bayesian analyses, one may have limited beliefs about a parameter and there may be several priors that provide suitable matches to these beliefs. In estimating a normal mean, we illustrate the use of two distinct priors in modeling beliefs, and show that inferences may or may not be sensitive to the choice of prior. In this example, we illustrate the "brute force" method of summarizing a posterior where the density is computed by the "prior times likelihood" recipe over a fine grid. We conclude by describing a Bayesian test of the simple hypothesis that a coin is fair. The computation of the posterior probability of "fair coin" is facilitated using `beta` and `binom` functions in R.

3.2 Normal Distribution with Known Mean but Unknown Variance

Gelman et al (2003) consider a problem of estimating an unknown variance using American football scores. The focus is on the difference d between a game outcome (winning score minus losing score) and a published point spread. We observe $d_1, ..., d_n$, the observed differences between game outcomes and point spreads for n football games. If these differences are assumed to be a random sample from a normal distribution with mean 0 and unknown variance σ^2, the likelihood function is given by

$$L(\sigma^2) = (\sigma^2)^{-n/2} \exp\{-\sum_{i=1}^{n} d_i^2/(2\sigma^2)\}, \ \sigma^2 > 0.$$

Suppose the noninformative prior density $p(\sigma^2) = 1/\sigma^2$ is assigned to the variance. Then the posterior density of σ^2 is given, up to a proportionality constant, by

$$g(\sigma^2|\text{data}) \propto (\sigma^2)^{-n/2-1} \exp\{-v/(2\sigma^2)\},$$

where $v = \sum_{i=1}^{n} d_i^2$. If we define the precision parameter $P = 1/\sigma^2$, then it can be shown that P is distributed as U/v, where U has a chi-squared distribution with n degrees of freedom. Suppose we are interested in a point estimate and a 95% probability interval for the standard deviation σ.

In the following R output, we first read in the datafile `footballscores` that is available in the LearnBayes package. For each of 672 games, the data file contains `favorite` and `underdog`, the actual scores of the favorite and underdog teams, and `spread`, the published point spread. We compute the difference variable d. As in the preceding notation, `n` is the sample size and `v` is the sum of squares of the differences.

```
> data(footballscores)
> attach(footballscores)
> d = favorite - underdog - spread
> n = length(d)
> v = sum(d^2)
```

We simulate 1000 values from the posterior distribution of the standard deviation σ in two steps. First, we simulate values of the precision parameter $P = 1/\sigma^2$ from the scaled chi-square(n) distribution by the command `rchisq(1000, n)/v`. Then we perform the transformation $\sigma = \sqrt{1/P}$ to get values from the posterior distribution of the standard deviation σ. We use the `hist` command to construct a histogram of the draws of σ (see Fig. 3.1).

```
> P = rchisq(1000, n)/v
> s = sqrt(1/P)
> hist(s,main="")
```

The R `quantile` command is used to extract the 2.5%, 50%, and 97.5% percentiles of this simulated sample. A point estimate for σ is provided by the posterior median 13.85 . In addition, the extreme percentiles (13.2, 14.6) represent a 95% probability interval for σ.

```
> quantile(s, probs = c(0.025, 0.5, 0.975))

    2.5%       50%     97.5%
13.17012  13.85135  14.56599
```

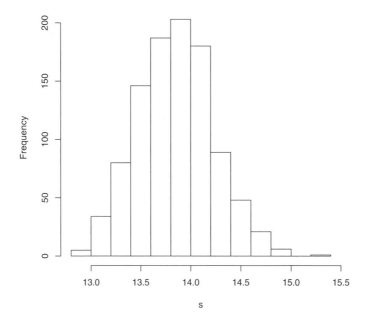

Fig. 3.1. Histogram of simulated sample of the standard deviation σ of differences between game outcomes and point spreads.

3.3 Estimating a Heart Transplant Mortality Rate

Consider the problem of learning about the rate of success of heart transplant surgery of a particular hospital in the United States. For this hospital, we observe the number of transplant surgeries n and the number of deaths within 30 days of surgery y is recorded. In addition, one can predict the probability of death for an individual patient. This prediction is based on a model that uses information such as patients' medical condition before surgery, gender, and race. Based on these predicted probabilities, one can obtain an expected number of deaths, denoted by e. A standard model assumes that the number of deaths y follows a Poisson distribution with mean $e\lambda$, and the objective is to estimate the mortality rate per unit exposure λ.

The standard estimate of λ is the maximum likelihood estimate $\hat{\lambda} = y/e$. Unfortunately, this estimate can be poor when the number of deaths y is close to zero. In this situation when small death counts are possible, it is desirable to use a Bayesian estimate that uses prior knowledge about the size of the mortality rate. A convenient choice for a prior distribution is a member of the gamma(α, β) density of the form

$$p(\lambda) \propto \lambda^{\alpha-1} \exp(-\beta\lambda), \ \lambda > 0.$$

A convenient source of prior information is heart transplant data from a small group of hospitals that we believe has the same rate of mortality as the rate from the hospital of interest. Suppose we observe the number of deaths z_j and the exposure o_j for 10 hospitals ($j = 1, ..., 10$), where z_j is Poisson with mean $o_j\lambda$. If we assign λ the standard noninformative prior $p(\lambda) = \lambda^{-1}$, then the updated distribution for λ, given this data from the 10 hospitals, is

$$p(\lambda) \propto \lambda^{\sum_{j=1}^{10} z_j - 1} \exp\Big(-\big(\sum_{j=1}^{10} o_j\big)\lambda\Big).$$

Using this information, we have a gamma(α, β) prior for λ, where $\alpha = \sum_{j=1}^{10} z_j$ and $\beta = \sum_{j=1}^{10} o_j$. In this example, we have

$$\sum_{j=1}^{10} z_j = 16, \sum_{j=1}^{10} o_j = 15174,$$

and so we assign λ a gamma(16, 15174) prior.

If the observed number of deaths from surgery y_{obs} for a given hospital with exposure e is Poisson $(e\lambda)$ and λ is assigned the gamma(α, β) prior, then the posterior distribution will also have the gamma form with parameters $\alpha + y_{\mathrm{obs}}$ and $\beta + e$. Also the (prior) predictive density of y (before any data are observed) can be computed by the formula

$$f(y) = \frac{f(y|\lambda)g(\lambda)}{g(\lambda|y)},$$

where $f(y|\lambda)$ is the Poisson$(e\lambda)$ sampling density and $g(\lambda)$ and $g(\lambda|y)$ are, respectively, the prior and posterior densities of λ.

By the model checking strategy of Box (1980), both the posterior density $g(\lambda|y)$ and the predictive density $f(y)$ play important roles in a Bayesian analysis. By use of the posterior density, one performs inference about the unknown parameter conditional on the Bayesian model that includes the assumptions of sampling density and the prior density. One can check the validity of the proposed model by inspecting the predictive density. If the observed data value y_{obs} is consistent with the predictive density $p(y)$, then the model seems reasonable. On the other hand, if y_{obs} is in the extreme tail portion of the predictive density, then this casts doubt on the validity of the Bayesian model, and perhaps the prior density or the sampling density has been misspecified.

We consider inference about the heart transplant rate for two hospitals – one that has experienced a small number of surgeries and a second that has experienced many surgeries. First consider hospital A which experienced only one death ($y_{\mathrm{obs}} = 1$) with an exposure of $e = 66$. The standard estimate of this hospital's rate, 1/66, is suspect due to the small observed number of deaths. The following R calculations illustrate the Bayesian calculations. After the gamma prior parameters `alpha` and `beta` and exposure `ex` are defined, the predictive density of the values $y = 0, 1, ..., 10$ are found by use of the preceding formula and the R functions `dpois` and `dgamma`. The formula for the predictive density is valid for all λ, but to ensure that there is not any underflow in the calculations, the values of $f(y)$ are computed for the prior mean value $\lambda = \alpha/\beta$. Note that practically all of the probability of the predictive density is concentrated on the two values $y = 0$ and 1. The observed number of deaths ($y_{obs} = 1$) is in the middle of this predictive distribution and so there is no reason to doubt our Bayesian model.

```
> alpha=16;beta=15174
> yobs=1; ex=66
> y=0:10
> lam=alpha/beta
> py=dpois(y, lam*ex)*dgamma(lam, shape = alpha,
+    rate = beta)/dgamma(lam, shape= alpha + y,
+    rate = beta + ex)
> cbind(y, round(py, 3))

          y
 [1,]  0 0.933
 [2,]  1 0.065
 [3,]  2 0.002
 [4,]  3 0.000
 [5,]  4 0.000
 [6,]  5 0.000
 [7,]  6 0.000
 [8,]  7 0.000
 [9,]  8 0.000
[10,]  9 0.000
[11,] 10 0.000
```

The posterior density of λ can be summarized by simulating 1000 values from the gamma density.

```
> lambdaA = rgamma(1000, shape = alpha + yobs, rate = beta + ex)
```

Let's consider the estimation of a different hospital that experiences many surgeries. Hospital B had $y_{obs} = 4$ deaths with an exposure of $e = 1767$. For this data, we again have R compute the prior predictive density and simulate 1000 draws from the posterior density using the `rgamma` command. Again we see that the observed number of deaths seems consistent with this model since $y_{obs} = 4$ is not in the extreme tails of this distribution.

```
> ex = 1767; yobs=4
> y = 0:10
```

```
> py = dpois(y, lam * ex) * dgamma(lam, shape = alpha,
+       rate = beta)/dgamma(lam, shape = alpha + y,
+       rate = beta + ex)
> cbind(y, round(py, 3))

          y
 [1,]   0 0.172
 [2,]   1 0.286
 [3,]   2 0.254
 [4,]   3 0.159
 [5,]   4 0.079
 [6,]   5 0.033
 [7,]   6 0.012
 [8,]   7 0.004
 [9,]   8 0.001
[10,]   9 0.000
[11,]  10 0.000

> lambdaB = rgamma(1000, shape = alpha + yobs, rate = beta + ex)
```

To see the impact of the prior density on the inference, it is helpful to display the prior and posterior distributions on the same graph. In Fig. 3.2, the histograms of the draws from the posterior distributions of the rates are shown for hospitals A and B. The gamma prior density is placed on top of the histogram in each case. We see that for hospital A with relatively little experience in surgeries, the prior information is significant and the posterior distribution resembles the prior distribution. In contrast, for hospital B with many surgeries, the prior information is less influential and the posterior distribution resembles the likelihood function.

```
> lambda = seq(0, max(c(lambdaA,lambdaB)), length = 500)
> par(mfrow = c(2, 1))
> hist(lambdaA, freq = FALSE, main="", ylim=c(0,1500))
> lines(lambda, dgamma(lambda, shape = alpha, rate = beta))
> hist(lambdaB, freq = FALSE, main="", ylim=c(0,1500))
> lines(lambda, dgamma(lambda, shape = alpha, rate = beta))
```

3.4 An Illustration of Bayesian Robustness

In practice, one may have incomplete prior information about a parameter in the sense that one's beliefs won't entirely define a prior density. There may be a number of different priors that match the given prior information. For example, if you believe a priori that the median of a parameter θ is 30 and its 80th percentile is 50, certainly there are many prior probability distributions that can be chosen that match these two percentiles. In this situation where

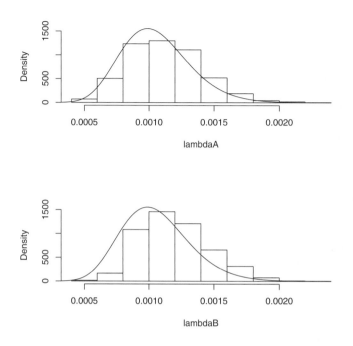

Fig. 3.2. Histograms for simulated samples from the posterior distributions for two transplant rates. The prior density for the corresponding rate is drawn in each graph.

different priors are possible, it is desirable that inferences from the posterior not be dependent on the exact functional form of the prior. A Bayesian analysis is said to be *robust* to the choice of prior if the inference is insensitive to different priors that match the user's beliefs.

To illustrate this idea, suppose you are interested in estimating the true IQ θ for a person we'll call Joe. You believe Joe has an average intelligence and the median of your prior distribution is 100. Also you are 90% confident that Joe's IQ falls between 80 and 120. You construct a prior density by matching this information with a normal density with mean μ and standard deviation τ, or $N(\mu, \tau)$. It is straightforward to show that the parameter values that match this prior information are $\mu = 100$ and $\tau = 12.16$.

Joe takes four IQ tests and his scores are y_1, y_2, y_3, y_4. Assuming that an individual score y is distributed $N(\theta, \sigma)$ with known standard deviation $\sigma = 15$, the observed mean score $\bar{y}$ is $N(\theta, \sigma/\sqrt{4})$.

With the use of a normal prior in this case, the posterior density of θ will also be normal with standard deviation

$$\tau_1 = 1/(\sqrt{4/\sigma^2 + 1/\tau^2})$$

and mean
$$\mu_1 = \frac{\bar{y}(4/\sigma^2) + \mu(1/\tau^2)}{4/\sigma^2 + 1/\tau^2}.$$

We illustrate the posterior calculations for three hypothetical test results for Joe. We suppose that the observed mean test score is $\bar{y} = 110$, or $\bar{y} = 125$, or $\bar{y} = 140$. In each case we compute the posterior mean and posterior standard deviation of Joe's true IQ θ. These values are denoted by the R variables mu1 and tau1 in the following output.

```
> mu = 100
> tau = 12.16
> sigma = 15
> n = 4
> se = sigma/sqrt(4)
> ybar = c(110, 125, 140)
> tau1 = 1/sqrt(1/se^2 + 1/tau^2)
> mu1 = (ybar/se^2 + mu/tau^2) * tau1^2
> summ1=cbind(ybar, mu1, tau1)
> summ1
```

```
      ybar       mu1      tau1
[1,]   110 107.2442 6.383469
[2,]   125 118.1105 6.383469
[3,]   140 128.9768 6.383469
```

Let's now consider an alternative prior density to model our beliefs about Joe's true IQ. Any symmetric density instead of a normal could be used, so we use a t density with location μ, scale τ, and two degrees of freedom. Since our prior median is 100, we let the median of our t density be equal to $\mu = 100$. We find the scale parameter τ so the t density matches our prior belief that the 95th percentile of θ is equal to 120. Note that

$$P(\theta < 120) = P(T < \frac{20}{\tau}) = .95,$$

where T is a standard t variate with two degrees of freedom. It follows that

$$\tau = 20/t_2(.95),$$

where $t_v(p)$ is the pth quantile of a t random variable with v degrees of freedom. We find τ by use of the t quantile function qt in R.

```
> tscale = 20/qt(0.95, 2)
> tscale
```

```
[1] 6.849349
```

We display the normal and t priors in a single graph in Fig. 3.3. Although they have the same basic shape, note that the t density has significantly flatter tails – we will see that this will impact the posterior density for "extreme" test scores.

```
> theta = seq(60, 140, length = 200)
> plot(theta,1/tscale*dt((theta-mu)/tscale,2),
+   type="l",ylab="Prior Density")
> lines(theta,1/10*dnorm((theta-mu)/tau),lwd=3)
> legend(locator(1),legend=c("t density","normal density"),
+   lwd=c(1,3))
```

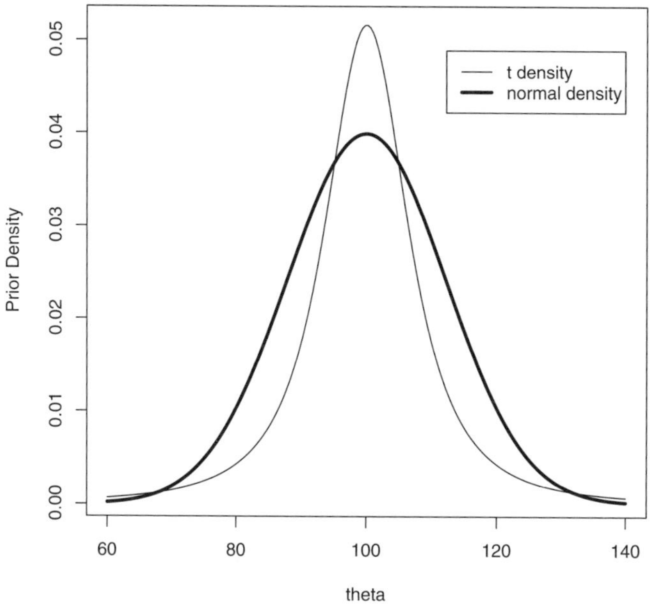

Fig. 3.3. Normal and t priors for representing prior opinion about a person's true IQ score.

We perform the posterior calculations using the t prior for each of the possible sample results. Note that the posterior density of θ is given, up to a proportionality constant, by

$$g(\theta|\text{data}) \propto \phi(\bar{y}|\theta, \sigma/\sqrt{n})g_T(\theta|v, \mu, \tau),$$

where $\phi(y|\theta, \sigma)$ is a normal density with mean θ and standard deviation σ, and $g_T(\mu|v, \mu, \tau)$ is a t density with median μ, scale parameter τ and degrees of freedom v. Since this density does not have a convenient functional form, we summarize it by a direct "prior times likelihood" approach. We construct a grid of θ values that covers the posterior density, compute the product of the normal likelihood and the t prior on the grid, and convert these products to probabilities by dividing by the sum. Essentially we are approximating the continuous posterior density by a discrete distribution on this grid. We then use this discrete distribution to compute the posterior mean and posterior standard deviation. We apply this computation algorithm for the three values of $\bar{y}$ and the posterior moments are displayed in the second and third columns of the R matrix summ2.

```
> summ2 = c()
> for (i in 1:3) {
+      theta = seq(60, 180, length = 500)
+      like = dnorm((theta - ybar[i])/7.5)
+      prior = dt((theta - mu)/tscale, 2)
+      post = prior * like
+      post = post/sum(post)
+      m = sum(theta * post)
+      s = sqrt(sum(theta^2 * post) - m^2)
+      summ2 = rbind(summ2, c(ybar[i], m, s))
+ }
> summ2

      [,1]     [,2]      [,3]
[1,]  110 105.2921 5.841676
[2,]  125 118.0841 7.885174
[3,]  140 135.4134 7.973498
```

Let's compare the posterior moments of θ using the two priors by combining the two R matrices summ1 and summ2.

```
> cbind(summ1,summ2)

      [,1]     [,2]      [,3] [,4]     [,5]      [,6]
[1,]  110 107.2442 6.383469  110 105.2921 5.841676
[2,]  125 118.1105 6.383469  125 118.0841 7.885174
[3,]  140 128.9768 6.383469  140 135.4134 7.973498
```

When $\bar{y} = 110$, the values of the posterior mean and posterior standard deviation are similar using the normal and t priors. However, there can be substantial differences in the posterior moments using the two priors when the observed mean score is inconsistent with the prior mean. In the "extreme" case where $\bar{y} = 140$, Fig. 3.4 graphs the posterior densities for the two priors.

```
> normpost = dnorm(theta, mu1[3], tau1)
> normpost = normpost/sum(normpost)
> plot(theta,normpost,type="l",lwd=3,ylab="Posterior Density")
> lines(theta,post)
> legend(locator(1),legend=c("t prior","normal prior"),lwd=c(1,3))
```

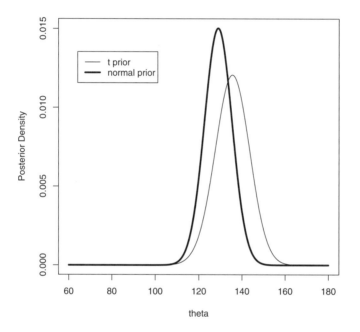

Fig. 3.4. Posterior densities for a person's true IQ using normal and t priors.

When a normal prior is used, the posterior will always be a compromise between the prior information and the observed data, even when the data result
conflicts with one's prior beliefs about the location of Joe's IQ. In contrast,
when a t prior is used, the likelihood will be in the flat-tailed portion of the
prior and the posterior will resemble the likelihood function.

In this case, the inference about the mean is robust to the choice of prior
(normal or t) when the observed mean IQ score is consistent with the prior
beliefs. But in the case when an extreme IQ score is observed, we see that the
inference is not robust to the choice of prior density.

3.5 A Bayesian Test of the Fairness of a Coin

Suppose you are interested in assessing the fairness of a coin. You observe y distributed binomial with parameters n and p and you are interested in testing the hypothesis H that $p = .5$. If y is observed, then it is usual practice to make a decision on the basis of the p-value

$$2 \times P(Y \leq y).$$

If this p-value is *small*, then you reject the hypothesis H and conclude that the coin is not fair. Suppose, for example, the coin is flipped 20 times and only 5 heads are observed. In R we compute the probability of obtaining five or fewer heads:

```
> pbinom(5, 20, 0.5)

[1] 0.02069473
```

The p-value here is $2 \times .021 = .042$. Since this value is smaller than the common significance level of .05, you would decide to reject the hypothesis H and conclude that the coin is not fair.

Let's consider this problem from a Bayesian perspective. There are two possible models here – either the coin is fair ($p = .5$) or the coin is not fair ($p \neq .5$). Suppose that you are indifferent between the two possibilities, and so you initially assign each model a probability of $1/2$. Now if you believe the coin is fair, then your entire prior distribution for p would be concentrated on the value $p = .5$. If instead the coin is unfair, you would assign a different prior distribution on $(0, 1)$, call it $g_1(p)$, that would reflect your beliefs about the unfair coin probability. Suppose you assign a beta(a, a) prior on p. This beta distribution is symmetric about .5 – it says that you believe the coin is not fair, the probability is close to $p = .5$. To summarize, your prior distribution in this testing situation can be written

$$g(p) = .5I(p = .5) + .5I(p \neq .5)g_1(p),$$

where $I(A)$ is an indicator function equal to 1 if the event A is true and otherwise equal to 0.

After observing the number of heads in n tosses, we would update our prior distribution by Bayes' rule. The posterior density for p can be written as

$$g(p|y) = \lambda(y)I(p = .5) + (1 - \lambda(y))g_1(p|y),$$

where g_1 is a beta$(a+y, a+n-y)$ density and $\lambda(y)$ is the posterior probability of the model that the coin is fair

$$\lambda(y) = \frac{.5p(y|.5)}{.5p(y|.5) + .5m_1(y)}.$$

In the expression for $\lambda(y)$, $p(y|.5)$ is the binomial density for y when $p = .5$, and $m_1(y)$ is the (prior) predictive density for y using the beta density.

In R the posterior probability of fairness $\lambda(y)$ is easily computed. The R command dbinom will compute the binomial probability $p(y|.5)$ and the predictive density for y can be computed using the identity

$$m_1(y) = \frac{f(y|p)g_1(p)}{g_1(p|y)}.$$

Assume first that we assign a beta(10, 10) prior for p when the coin is not fair and we observe $y = 5$ heads in $n = 20$ tosses. The posterior probability of fairness is stored in the R variable lambda.

```
> n = 20
> y = 5
> a = 10
> p = 0.5
> m1 = dbinom(y, n, p) * dbeta(p, a, a)/dbeta(p, a + y, a + n -
+      y)
> lambda = dbinom(y, n, p)/(dbinom(y, n, p) + m1)
> lambda
```

```
[1] 0.2802215
```

We get the surprising result that the posterior probability of the hypothesis of fairness H is .28, which is less evidence against fairness than implied by the above p-value calculation.

The function pbetat in the LearnBayes package performs a test of a binomial proportion. The inputs to the function are the value of p to be tested, the prior probability of that value, a vector of parameters of the beta prior when the hypothesis is not true, and a vector of numbers of successes and failures. In this example, the syntax would be

```
> pbetat(p,.5,c(a,a),c(y,n-y))
```

```
$bf
[1] 0.3893163
```

```
$post
[1] 0.2802215
```

The output variable post is the posterior probability that $p = .5$, which agrees with the calculation. The output variable bf is the Bayes factor in support of the null hypothesis which is discussed in Chapter 8.

Since the choice of the prior parameter $a = 10$ in this analysis seems arbitrary, it is natural to ask about the sensitivity of this posterior calculation to the choice of this parameter. To answer this question, we compute the posterior probability of fairness for a range of values of $\log a$. We graph the posterior probability against $\log a$ in Fig. 3.5.

```
> loga = seq(-4, 5, length = 100)
> a = exp(loga)
> m2 = dbinom(y, n, p) * dbeta(p, a, a)/dbeta(p, a + y, a + n -
+    y)
> lambda = dbinom(y, n, p)/(dbinom(y, n, p) + m2)

> plot(loga,lambda,type="l",xlab="log(a)",ylab="Prob(coin is fair)")
```

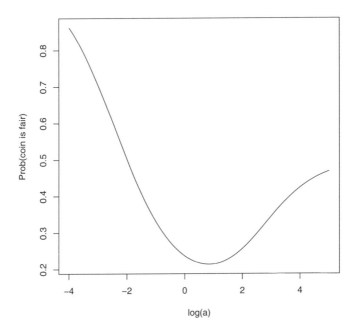

Fig. 3.5. Posterior probability that coin is fair graphed against values of the prior parameter $\log a$.

We see from this graph that the probability of fairness appears to be greater than .2 for all choices of a. Since the value of .2 is larger than the p-value calculation of .042, this suggests that the p-value is overstating the evidence against the hypothesis that the coin is fair.

Another distinction between the frequentist and Bayesian calculations is the event that led to the decision about rejecting the hypothesis that the coin was fair. The p-value calculation was based on the probability of the event "5 heads or fewer," but the Bayesian calculation was based solely on the likelihood based on the event "exactly 5 heads." That raises the question: how would the Bayesian answers change if we observed "5 heads or fewer"? One can show that the posterior probability that the coin is fair is given by

$$\lambda(y) = \frac{.5P_0(Y \leq 5)}{.5P_0(Y \leq 5) + .5P_1(Y \leq 5)}.$$

where $P_0(Y \leq 5)$ is the probability of five heads or fewer under the binomial model with $p = .5$, and $P_1(Y \leq 5)$ is the predictive probability of this event under the alternative model with a beta$(10, 10)$ prior on p. In the following R output, the cumulative probability of five heads under the binomial model is computed by the R function `pbinom`. The probability of five or fewer heads under the alternative model is computed by summing the predictive density over the six values of y.

```
> n=20
> y=5
> a=10
> p=.5
> m2=0
> for (k in 0:y)
+ {m2=m2+dbinom(k,n,p)*dbeta(p,a,a)/dbeta(p,a+k,a+n-k)}
> lambda=pbinom(y,n,p)/(pbinom(y,n,p)+m2)
> lambda
```

[1] 0.2184649

Note that the posterior probability of fairness is .218 based on the data "5 heads or fewer." This posterior probability is smaller than the value of .280 found earlier based on $y = 5$. This is a reasonable result, since observing "5 heads or fewer" is stronger evidence against fairness than the result "5 heads."

3.6 Further Reading

Carlin and Louis (2000), chapter 2, and Gelman et al (2003), chapter 2, provide general discussions of Bayesian inference for one-parameter problems. Lee (2004), Antleman (1996), and Bolstad (2004) provide extensive descriptions of inference for a variety of one-parameter models. The notion of Bayesian robustness is discussed in detail in Berger (1985). Bayesian testing for basic inference problems is outlined in Lee (2004).

3.7 Summary of R Functions

`pbetat` – Bayesian test that a proportion is equal to a specified prior using a beta prior

Usage: pbetat(p0,prob,ab,data)

Arguments: p0, the value of the proportion to be tested; prob, the prior probability of the hypothesis; ab, the vector of parameter values of the beta prior under the alternative hypothesis; data, vector containing the number of successes and number of failures

Value: bf, the Bayes factor in support of the null hypothesis; post, the posterior probability of the null hypothesis

3.8 Exercises

1. **Cauchy sampling model**

   Suppose one observes a random sample $y_1, ..., y_n$ from a Cauchy density with location θ and scale parameter 1. If a uniform prior is placed on θ, then the posterior density is given (up to a proportionality constant) by

   $$g(\theta|data) \propto \prod_{i=1}^{n} \frac{1}{1 + (y_i - \theta)^2}.$$

   Suppose one observes the data 0, 10, 9, 8, 11, 3, 3, 8, 8, 11.
   a) Using the R command seq, set up a grid of values of θ from -2 to 12 in steps of 0.1.
   b) Compute the posterior density on this grid.
   c) Plot the density and comment on its main features.
   d) Compute the posterior mean and posterior standard deviation of θ.

2. **Learning about an exponential mean**

   Suppose a random sample is taken from an exponential distribution with mean λ. If we assign the usual noninformative prior $g(\lambda) = 1/\lambda$, then the posterior density is given, up to a proportionality constant, by

   $$g(\lambda|data) = \lambda^{-n-1} \exp\{-s/\lambda\},$$

   where n is the sample size and s is the sum of the observations.
   a) Show that if we transform λ to $\theta = 1/\lambda$, then λ has a gamma density with shape parameter n and rate parameter s. (A gamma density with shape α and rate β is proportional to $h(x) = x^{\alpha-1} \exp(-\beta x)$.)
   b) In a life-testing illustration, five bulbs are tested with observed burn times (in hours) of 751, 594, 1213, 1126, and 819. Using the R function rgamma, simulate 1000 values from the posterior distribution of θ.
   c) By transforming these simulated draws, obtain a simulated sample from the posterior distribution of λ.
   d) Estimate the posterior probability that λ exceeds 1000 hours.

3. **Learning about the upper bound of a discrete uniform density**

   Suppose one takes independent observations $y_1, ..., y_n$ from a uniform distribution on the set $\{1, 2, ..., N\}$, where the upper bound N is unknown.

Suppose one places a uniform prior for N on the values $1, ..., B$, where B is known. Then the posterior probabilities for N are given by

$$g(N|y) \propto \frac{1}{N^n}, \; y_{(n)} \leq N \leq B,$$

where $y_{(n)}$ is the maximum observation. To illustrate this situation, suppose a tourist is waiting for a taxi in a city. During this waiting time, she observes five taxis with the numbers 43, 24, 100, 35, and 85. She assumes that taxis in this city are numbered from 1 to N, she is equally likely to observe any numbered taxi at a given time, and observations are independent. She also knows that there cannot be more than 200 taxis in the city.

a) Use R to compute the posterior probabilities of N on a grid of values.
b) Compute the posterior mean and posterior standard deviation of N.
c) Find the probability that there are more than 150 taxis in the city.

4. **Bayesian robustness**
Suppose you are about to flip a coin that you believe is fair. If p denotes the probability of flipping a head, then your "best guess" at p is .5. Moreover, you believe that it is highly likely that the coin is close to fair, which you quantify by $P(.44 < p < .56) = .9$. Consider the following two priors for p:
P1:p distributed beta(100, 100)
P2:p distributed according to the mixture prior

$$g(p) = .9f_B(p; 500, 500) + .1f_B(p; 1, 1),$$

where $f_B(p; a, b)$ is the beta density with parameters a and b.
a) Simulate 1000 values from each prior density P1 and P2. By summarizing the simulated samples, show that both priors match the given prior beliefs about the coin flipping probability p.
b) Suppose you flip the coin 100 times and obtain 45 heads. Simulate 1000 values from the posteriors from priors P1 and P2, and compute 90% probability intervals.
c) Suppose that you only observe 30 heads out of 100 flips. Again simulate 1000 values from the two posteriors and compute 90% probability intervals.
d) Looking at your results from (b) and (c), comment on the robustness of the inference with respect to the choice of prior density in each case.

5. **Test of a proportion**
In the standard Rhine test of extra-sensory perception (ESP), a set of cards is used where each card has a circle, a square, wavy lines, a cross, or a star. A card is selected at random from the deck and a person tries to guess the symbol on the card. This experiment is repeated 20 times and the number of correct guesses y is recorded. Let p denote the probability that the person makes a correct guess, where $p = .2$ if the person does not

have ESP and is just guessing at the card symbol. To see if the person truly has some ESP, we would like to test the hypothesis $H : p = .2$.

a) If the person identifies $y = 8$ cards correctly, compute the p-value.
b) Suppose you believe a priori that the probability that $p = .2$ is .5 and if $p \neq .2$, you assign a beta(1, 4) prior on the proportion. Using the function pbetat, compute the posterior probability of the hypothesis H. Compare your answer with the p-value computed in part (a).
c) The posterior probability computed in part (b) depended on your belief about plausible values of the proportion p when $p \neq .2$. For each of the following priors, compute the posterior probability of H:
   (1) $p \sim$ beta(.5, 2), (2) $p \sim$ beta(2, 8), (3) $p \sim$ beta(8, 32).
d) On the basis of your Bayesian computations, do you think that $y = 8$ is convincing evidence that the person really has some ESP? Explain.

6. **Learning from grouped data**
   Suppose you drive on a particular interstate roadway and typically drive at a constant speed of 70 mph. One day, you pass one car and get passed by 17 cars. Suppose that the speeds are normally distributed with unknown mean μ and standard deviation $\sigma = 10$. If you pass s cars, and f cars pass you, the likelihood of μ is given by

   $$L(\mu) = \Phi(70, \mu, \sigma)^s (1 - \Phi(70, \mu, \sigma))^f,$$

   where $\Phi(y, \mu, \sigma)$ is the cdf of the normal distribution with mean μ and standard deviation σ. Assign the unknown mean μ a flat prior density.
   a) If $s = 1$ and $f = 17$, plot the posterior density of μ.
   b) Using the density found in part (a), find the posterior mean of μ.
   c) Find the probability that the average speed of the cars on this interstate roadway exceeds 80 mph.

4

Multiparameter Models

4.1 Introduction

In this chapter, we describe the use of R to summarize Bayesian models with several unknown parameters. In learning about parameters of a normal population or multinomial parameters, posterior inference is accomplished by simulating from distributions of standard forms. Once a simulated sample is obtained from the joint posterior, it is straightforward to perform transformations on these simulated draws to learn about any function of the parameters. We next consider estimating the parameters of a simple logistic regression model. Although the posterior distribution does not have a simple functional form, it can be summarized by computing the density on a fine grid of points. A common inference problem is to compare two proportions in a 2×2 contingency table. We illustrate the computation of the posterior probability that one proportion exceeds the second proportion in the situation in which one believes a priori that the proportions are dependent.

4.2 Normal Data with Both Parameters Unknown

A standard inference problem is to learn about a normal population where both the mean and variance are unknown. To illustrate Bayesian computation for this problem, suppose we are interested in learning about the distribution of completion times for men between ages 20 and 29 who are running the New York Marathon. We observe the times $y_1, \ldots, y_{20}$ for 20 runners in minutes, and we assume they represent a random sample from an $N(\mu, \sigma)$ distribution. If we assume the standard noninformative prior $g(\mu, \sigma^2) \propto 1/\sigma^2$, then the posterior density of the mean and variance is given by

$$g(\mu, \sigma^2 | y) \propto \frac{1}{(\sigma^2)^{n/2+1}} \exp(-\frac{1}{2\sigma^2}(S + n(\mu - \bar{y})^2)),$$

where n is the sample size, $\bar{y}$ is the sample mean, and $S = \sum_{i=1}^{n}(y_i - \bar{y})^2$.

This joint posterior has the familiar normal/inverse chi-square form where

- the posterior of μ conditional on σ^2 is distributed $N(\bar{y}, \sigma/\sqrt{n})$
- the marginal posterior of σ^2 is distributed $S\chi_{n-1}^{-2}$ where χ_{ν}^{-2} denotes an inverse chi-square distribution with ν degrees of freedom.

We first use R to construct a contour plot of the joint posterior density for this example. We read in the data `marathontimes`; when we `attach` this dataset, we can use the variable `time` that contains the vector of running times. The R function `normchi2post.R` in the LearnBayes package computes the logarithm of the joint posterior density of (μ, σ^2). We also use a function `mycontour.R` in the LearnBayes package that facilitates the use of the R `contour` command. There are three inputs to `mycontour`: the name of the function that defines the log density, a vector with the values (xlo, xhi, ylo, and yhi) that define the rectangle where the density is to graphed, and the data used in the function for the log density. The function produces a contour graph shown in Fig. 4.1, where the contour lines are drawn at 10%, 1%, and .1% of the maximum value of the posterior density over the grid.

```
> data(marathontimes)
> attach(marathontimes)
> d = mycontour(normchi2post, c(220, 330, 500, 9000), time)
> title(xlab="mean",ylab="variance")
```

It is convenient to summarize this posterior distribution by simulation. One can simulate a value of (μ, σ^2) from the joint posterior by first simulating σ^2 from an $S\chi_{n-1}^{-2}$ distribution and then simulating μ from the $N(\bar{y}, \sigma/\sqrt{n})$ distribution. In the following R output, we first simulate a sample of size 1000 from the chi-squared distribution by use of the function `rchisq`. Then simulated draws of the "scale times inverse chi-square" distribution of the variance σ^2 are obtained by transforming the chi-square draws. Finally, simulated draws of the mean μ are obtained by use of the function `rnorm`.

```
> S = sum((time - mean(time))^2)
> n = length(time)
> sigma2 = S/rchisq(1000, n - 1)
> mu = rnorm(1000, mean = mean(time), sd = sqrt(sigma2)/sqrt(n))
```

We display the simulated sampled values of (μ, σ^2) on top of the contour plot of the distribution in Fig. 4.1.

```
> points(mu, sigma2)
```

Inferences about the parameters or functions of the parameters are available from the simulated sample. To construct a 95% interval estimate for the mean μ, we use the R `quantile` function to find percentiles of the simulated sample of μ.

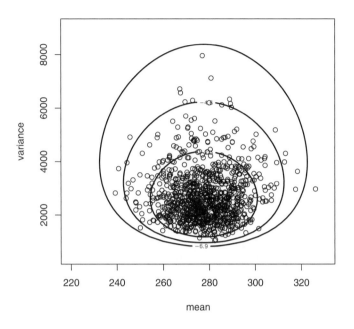

Fig. 4.1. Contour plot of the joint posterior distribution of (μ, σ^2) for a normal sampling model. The points represent a simulated random sample from this distribution.

```
> quantile(mu, c(0.025, 0.975))
     2.5%     97.5%
254.0937  301.7137
```

A 95% credible interval for the mean completion time is $(254.1, 301.7)$ minutes.

Suppose we are interested in learning about the standard deviation σ that describes the spread of the population of marathon running times. To obtain a sample of the posterior of σ, we take square roots of the simulated draws of σ^2. From the output, we see that an approximate 95% probability interval for σ is $(37.5, 70.9)$ minutes.

```
> quantile(sqrt(sigma2), c(0.025, 0.975))
     2.5%     97.5%
37.48217  70.89521
```

4.3 A Multinomial Model

Gelman et al (2003) describe a sample survey conducted by CBS news before the 1988 presidential election. A total of 1447 adults were polled to indicate their preference; $y_1 = 727$ supported George Bush, $y_2 = 583$ supported Michael Dukakis, and $y_3 = 137$ supported other candidates or expressed no opinion. The counts y_1, y_2, and y_3 are assumed to have a multinomial distribution with sample size n and respective probabilities θ_1, θ_2, and θ_3. If a uniform prior distribution is assigned to the multinomial vector $\theta = (\theta_1, \theta_2, \theta_3)$, then the posterior distribution of θ is proportional to

$$g(\theta) = \theta_1^{y_1} \theta_2^{y_2} \theta_3^{y_3},$$

which is recognized as a Dirichlet distribution with parameters $(y_1 + 1, y_2 + 1, y_3 + 1)$. The focus is to compare the proportions of voters for Bush and Dukakis by the difference $\theta_1 - \theta_2$.

The summarization of the Dirichlet posterior distribution is again conveniently done by simulation. Although the base R package does not have a function to simulate Dirichlet variates, it is easy to write a function to simulate this distribution based on the fact that if W_1, W_2, W_3 are independent distributed from gamma$(\alpha_1, 1)$, gamma$(\alpha_2, 1)$, gamma$(\alpha_3, 1)$ distributions and $T = W_1 + W_2 + W_3$, then the distribution of the proportions $(W_1/T, W_2/T, W_3/T)$ has a Dirichlet$(\alpha_1, \alpha_2, \alpha_3)$ distribution. The R function rdirichlet.R in the package LearnBayes uses this transformation of random variates to simulate draws of a Dirichlet distribution. One thousand vectors θ are simulated and stored in the matrix theta.

```
> alpha = c(728, 584, 138)
> theta = rdirichlet(1000, alpha)
```

Since we are interested in comparing the proportions for Bush and Dukakis, we focus on the difference $\theta_1 - \theta_2$. A histogram of the simulated draws of this difference is displayed in Fig. 4.2. Note that all of the mass of this distribution is on positive values, indicating that there is strong evidence that the proportion of voters for Bush exceeds the proportion for Dukakis.

```
> hist(theta[, 1] - theta[, 2], main="")
```

4.4 A Bioassay Experiment

In the development of drugs, bioassay experiments are often performed on animals. In a typical experiment, various dose levels of a compound are administered to batches of animals and a binary outcome (positive or negative) is recorded for each animal. We consider data from Gelman et al (2003), where

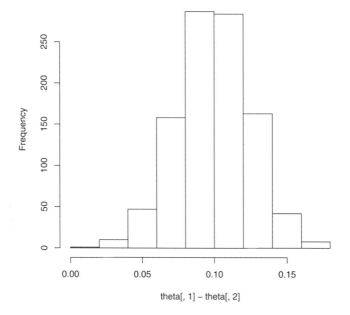

Fig. 4.2. Histogram of simulated sample of the marginal posterior distribution of $\theta_1 - \theta_2$ for the multinomial sampling example.

Table 4.1. Data from the bioassay experiment

Dose	Deaths	Sample size
−0.86	0	5
−0.30	1	5
−0.05	3	5
0.73	5	5

one observes a dose level (in log g/ml), the number of animals, and the number of deaths for each of four groups. The data are displayed in Table 4.1.

Let y_i denote the number of deaths observed out of n_i with dose level x_i. We assume y_i is binomial(n_i, p_i), where the probability p_i follows the logistic model

$$\log(p_i/(1 - p_i)) = \beta_0 + \beta_1 x_i.$$

The likelihood function of the unknown regression parameters β_0 and β_1 is given by

$$L(\beta_0, \beta_1) = \prod_{i=1}^{4} p_i^{y_i}(1 - p_i)^{n_i - y_i},$$

where $p_i = \exp(\beta_0 + \beta_1 x_i)/(1 + \exp(\beta_0 + \beta_1 x_i))$. If the standard flat noninformative prior is placed on (β_0, β_1), then the posterior density is proportional to the likelihood function.

We begin in R by defining the covariate vector x and the vectors of sample sizes and observed success counts n and y.

```
> x = c(-0.86, -0.3, -0.05, 0.73)
> n = c(5, 5, 5, 5)
> y = c(0, 1, 3, 5)
> data = cbind(x, n, y)
```

A standard classical analysis fits the model by maximum likelihood. The R function glm is used to do this fitting, and the summary output presents the estimates and the associated standard errors.

```
> response = cbind(y, n - y)
> results = glm(response ~ x, family = binomial)
> summary(results)

Call:
glm(formula = glmdata ~ x, family = binomial)

Deviance Residuals:
        1         2         3         4
-0.17236   0.08133  -0.05869   0.12237

Coefficients:
             Estimate Std. Error z value Pr(>|z|)
(Intercept)    0.8466     1.0191   0.831    0.406
x              7.7488     4.8728   1.590    0.112

(Dispersion parameter for binomial family taken to be 1)

    Null deviance: 15.791412  on 3  degrees of freedom
Residual deviance:  0.054742  on 2  degrees of freedom
AIC: 7.9648

Number of Fisher Scoring iterations: 7
```

The log posterior density for (β_0, β_1) in this logistic model is contained in the R function logisticpost. To summarize the posterior distribution, we first find a rectangle that covers essentially all of the posterior probability. The maximum likelihood fit is helpful for finding this rectangle. As seen by the contour plot displayed in Fig. 4.3, we see the rectangle $-4 \leq \beta_0 \leq 8, -5 \leq \beta_1 \leq 39$ contains the contours that are greater than .1% of the modal value.

```
> mycontour(logisticpost, c(-4, 8, -5, 39), data)
> title(xlab="beta0",ylab="beta1")
```

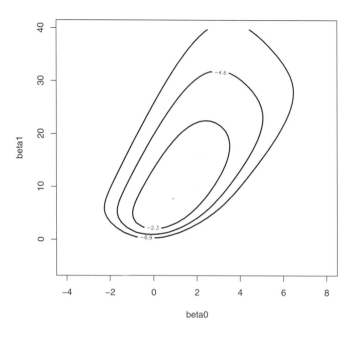

Fig. 4.3. Contour plot of the posterior distribution of (β_0, β_1) for the bioassay example. The contour lines are drawn at 10%, 1%, and .1% of the model height.

Now that we have found the posterior distribution, we use the function `simcontour` to simulate pairs of (β_0, β_1) from the posterior density computed on this rectangular grid. We display the contour plot with the points superimposed in Fig. 4.4 to confirm that we are sampling from the posterior distribution.

```
> s = simcontour(logisticpost, c(-4, 8, -5, 39), data, 1000)
> points(s$x, s$y)
```

We illustrate several types of inferences for this problem. Fig. 4.5 displays a density estimate of the simulated values (using the R function `density`) of the slope parameter β_1. All of the mass of the density of β_1 is on positive values, indicating that there is significant evidence that increasing the level of the dose does increase the probability of death.

```
> plot(density(s$y),xlab="beta1")
```

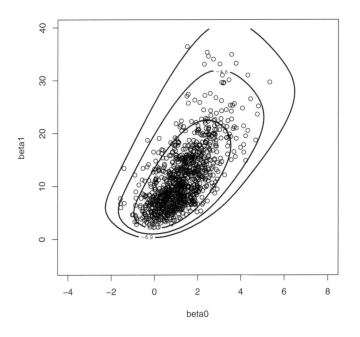

Fig. 4.4. Contour plot of the posterior distribution of (β_0, β_1) for the bioassay example. A simulated random sample from this distribution is shown on top of the contour plot.

In this setting, one parameter of interest is the LD-50, the value of the dose x such that the probability of death is equal to one half. It is straightforward to show that the LD-50 is equal to $\theta = -\beta_0/\beta_1$. One can obtain a simulated sample from the marginal posterior density of θ by computing a value of θ from each simulated pair (β_0, β_1). A histogram of the LD-50 is shown in Fig. 4.6.

```
> theta=-s$x/s$y
> hist(theta,xlab="LD-50")
```

In contrast to the histogram of β_1, the LD-50 is more difficult to estimate and the posterior density of this parameter is relatively wide. We compute a 95% credible interval from the simulated draws of θ.

```
> quantile(theta,c(.025,.975))
```

```
      2.5%       97.5%
-0.2903822   0.1140151
```

The probability that θ is contained in the interval $(-.290, .114)$ is .95.

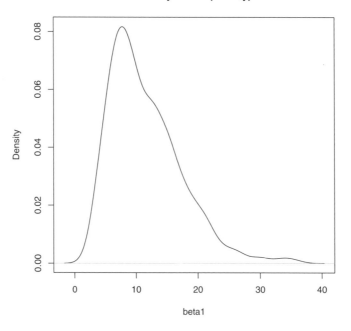

Fig. 4.5. Histogram of simulated values from the posterior of the slope parameter β_1 in the bioassay example.

4.5 Comparing Two Proportions

Howard (1998) considers the general problem of comparing the proportions from two independent binomial distributions. Suppose we observe y_1 distributed binomial(n_1, p_1), y_2 distributed binomial(n_2, p_2). One wants to know if the data favor the hypothesis $H_1 : p_1 > p_2$ or the hypothesis $H_2 : p_1 < p_2$ and want a measure of the strength of the evidence in support of one hypothesis. Howard gives a broad survey of frequentist and Bayesian approaches for comparing two proportions. Here we focus on the application of Howard's recommended "dependent prior" for this particular testing problem.

In this situation, suppose that one is given the information that one proportion is equal to a particular value, say $p_1 = .8$. This knowledge can influence a user's prior beliefs about the location of the second proportion p_2; specifically, one may believe that the value of p_2 is also close to .8. This implies that the use of dependent priors for p_1 and p_2 may be more appropriate than the common use of uniform independent priors for the proportions.

Howard proposes the following dependent prior. First the proportions are transformed into the real-valued logit parameters

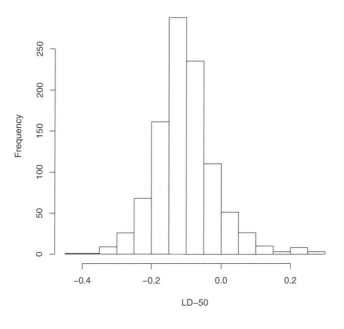

Fig. 4.6. Histogram of simulated values of the LD-50 parameter $-\beta_0/\beta_1$ in the bioassay example.

$$\theta_1 = \log \frac{p_1}{1-p_1}, \theta_2 = \log \frac{p_2}{1-p_2}.$$

Then suppose that given a value of θ_1, the logit θ_2 is assumed to be normally distributed with mean θ_1 and standard deviation σ. By generalizing this idea, Howard proposes the dependent prior of the general form

$$g(p_1, p_2) \propto e^{-(1/2)u^2} p_1^{\alpha-1}(1-p_1)^{\beta-1} p_2^{\gamma-1}(1-p_2)^{\delta-1}, 0 < p_1, p_2 < 1,$$

where

$$u = \frac{1}{\sigma} \log \frac{\theta_1}{\theta_2}.$$

This class of dependent priors is indexed by the parameters $(\alpha, \beta, \gamma, \delta, \sigma)$. The first four parameters reflect one's beliefs about the locations of p_1 and p_2 and the parameter σ indicates one prior belief in the dependence between the two proportions.

Suppose that $\alpha = \beta = \gamma = \delta = 1$, reflecting vague prior beliefs about each individual parameter. The logarithm of the dependent prior is defined in the R function `howardprior`. By use of the function `mycontour`, Fig. 4.7 shows contour plots of the dependent prior for values of the association parameter

$\sigma = 2, 1, .5,$ and $.25$. Note as the value of σ goes to zero, the prior is placing more of its mass along the line where the two proportions are equal.

```
> sigma=c(2,1,.5,.25)
> plo=.0001;phi=.9999
> par(mfrow=c(2,2))
> for (i in 1:4)
+ {
+ mycontour(howardprior,c(plo,phi,plo,phi),c(1,1,1,1,sigma[i]))
+ title(main=paste("sigma=",as.character(sigma[i])),
+   xlab="p1",ylab="p2")
+ }
```

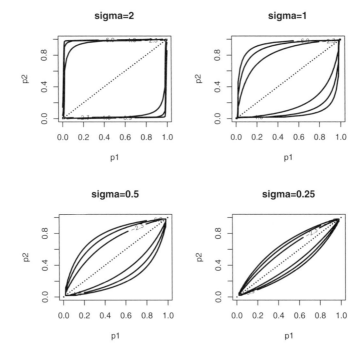

Fig. 4.7. Contour graphs of Howard's dependent prior for values of the association parameter $\sigma = 2, 1, .5,$ and $.25$.

Suppose we observe counts y_1, y_2 from the two binomial samples. The likelihood function is given by

$$L(p_1, p_2) = p_1^{y_1}(1 - p_1)^{n_1 - y_1} p_2^{y_2}(1 - p_2)^{n_2 - y_2}, 0 < p_1, p_2 < 1.$$

Combining the likelihood with the prior, one sees that the posterior density has the same functional "dependent" form with updated parameters

$$(\alpha + y_1, \beta + n_1 - y_1, \gamma + y_2, \delta + n_2 - y_2, \sigma).$$

We illustrate testing the hypotheses using a dataset discussed by Pearson (1947) shown in Table 4.2.

Table 4.2. Pearson's example

	Successes	Failures	Total
Sample 1	3	15	18
Sample 2	7	5	12
Totals	10	20	30

Since the posterior distribution is the same functional form as the prior, we can use the same **howardprior** function for the posterior calculations. In Fig. 4.8, contour plots of the posterior are shown for the four values of the association parameter σ.

```
> sigma=c(2,1,.5,.25)
> par(mfrow=c(2,2))
> for (i in 1:4)
+ {
+ mycontour(howardprior,c(plo,phi,plo,phi),
+   c(1+3,1+15,1+7,1+5,sigma[i]))
+ lines(c(0,1),c(0,1))
+ title(main=paste("sigma=",as.character(sigma[i])),
+   xlab="p1",ylab="p2")
+ }
```

We can test the hypothesis $H_1 : p_1 > p_2$ simply by computing the posterior probability of this region of the parameter space. We first produce by the function **simcontour** a simulated sample from the posterior distribution of (p_1, p_2) and then find the proportion of simulated pairs where $p_1 > p_2$. For example, we display the R commands for the computation of the posterior probability for $\sigma = 2$.

```
> s=simcontour(howardprior,c(plo,phi,plo,phi),
+   c(1+3,1+15,1+7,1+5,2),1000)
> sum(s$x>s$y)/1000
```

```
[1] 0.012
```

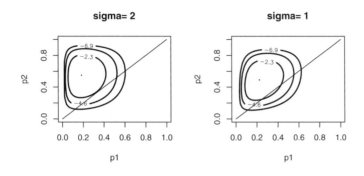

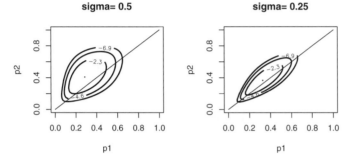

Fig. 4.8. Contour graphs of posterior for Howard's dependent prior for values of the association parameter $\sigma = 2$, 1, .5, and .25.

Table 4.3 displays the posterior probability that p_1 exceeds p_2 for four choices of the dependent prior parameter σ. Note that this posterior probability is sensitive to the prior belief about the dependence between the two proportions.

Table 4.3. Posterior probabilities of the hypothesis.

Dependent parameter σ	$P(p_1 > p_2)$
2	0.012
1	0.035
.5	0.102
.25	0.201

4.6 Further Reading

Chapter 3 of Gelman et al (2003) describes the normal sampling problem and other multiparameter problems from a Bayesian perspective. In particular, Gelman et al (2003) illustrate the use of simulation when the posterior has been computed on a grid. Carlin and Louis (2000), chapter 2, and Lee (2004) illustrate Bayesian inference for some basic two-parameter problems. Howard (1998) gives a general discussion of inference for the two-by-two contingency table, contrasting frequentist and Bayesian approaches.

4.7 Summary of R Functions

`howardprior` – computes the logarithm of a dependent prior on two proportions proposed by Howard in a *Statistical Science* paper in 1998
Usage: `howardprior(xy,par)`
Arguments: `xy`, a matrix of parameter values where each row represents a value of the proportions (p1, p2); `par`, a vector containing parameter values alpha, beta, gamma, delta, sigma
Value: vector of values of the log posterior where each value corresponds to each row of the parameters in xy

`logisticpost` – computes the log posterior density of (beta0, beta1) when yi are independent binomial(ni, pi) and logit(pi)=beta0+beta1*xi
Usage: `logisticpost(beta,data)`
Arguments: `beta`, a matrix of parameter values where each row represents a value of (beta0, beta1); `data`, a matrix of columns of covariate values x, sample sizes n, and number of successes y
Value: vector of values of the log posterior where each value corresponds to each row of the parameters in beta

`mycontour` – for a general two parameter density, draws a contour graph where the contour lines are drawn at 10%, 1%, and .1% of the height at the mode
Usage: `mycontour(logf,limits,data)`
Arguments: `logf`, a function that defines the logarithm of the density; `limits`, a vector of limits (xlo, xhi, ylo, yhi) where the graph is to be drawn; `data`, a vector or list of parameters associated with the function logpost
Value: a contour graph of the density is drawn

`normchi2post` – computes the log of the posterior density of a mean M and a variance S2 when a sample is taken from a normal density and a standard noninformative prior is used
Usage: `normchi2post(theta,data)`
Arguments: `theta`, a matrix of parameter values where each row is a value of (M, S2); `data`, a vector containing the sample observations
Value: a vector of values of the log posterior where the values correspond to the rows in theta

rdirichlet – simulates values from a Dirichlet distribution
Usage: rdirichlet(n,par)
Arguments: n, the number of simulations required; par, the vector of parameters of the Dirichlet distribution
Value: a matrix of simulated draws, where a row contains one simulated Dirichlet draw

simcontour – for a general two-parameter density defined on a grid, simulates a random sample
Usage: simcontour(logf,limits,data,m)
Arguments: logf, a function that defines the logarithm of the density; limits, a vector limits (xlo, xhi, ylo, yhi) that cover the joint probability density; data, a vector or list of parameters associated with the function logpost; m, the size of simulated sample
Value: x, the vector of simulated draws of the first parameter; y, the vector of simulated draws of the second parameter

4.8 Exercises

1. **Inference about a normal population**
   Suppose we are interested in learning about the sleeping habits of students at a particular college. We collect $y_1, ..., y_{20}$, the sleeping times (in hours), for 20 randomly selected students in a statistics course. Here are the observations:

9.0	8.5	7.0	8.5	6.0	12.5	6.0	9.0	8.5	7.5
8.0	6.0	9.0	8.0	7.0	10.0	9.0	7.5	5.0	6.5

   a) Assuming that the observations represent a random sample from a normal population with mean μ and variance σ^2 and the usual noninformative prior is placed on (μ, σ^2), simulate a sample of 1000 draws from the joint posterior distribution.
   b) Use the simulated sample to find 90% interval estimates for the mean μ and the standard deviation σ.
   c) Suppose one is interested in estimating the upper quartile p_{75} of the normal population. Using the fact that $p_{75} = \mu + 0.674\sigma$, find the posterior mean and posterior standard deviation of p_{75}.

2. **The Behrens-Fisher problem**
   Suppose that we observe two independent normal samples, the first distributed according to an $N(\mu_1, \sigma_1)$ distribution, the second according to an $N(\mu_2, \sigma_2)$ distribution. Denote the first sample by $x_1, ..., x_m$ and the second sample by $y_1, ..., y_n$. Suppose also that the parameters $(\mu_1, \sigma_1^2, \mu_2, \sigma_2^2)$ are assigned the vague prior

   $$g(\mu_1, \sigma_1^2, \mu_2, \sigma_2^2) \propto \frac{1}{\sigma_1^2 \sigma_2^2}.$$

a) Find the posterior density. Show that the vectors (μ_1, σ_1^2) and (μ_2, σ_2^2) have independent posterior distributions.

b) Describe how to simulate from the joint posterior density of $(\mu_1, \sigma_1^2, \mu_2, \sigma_2^2)$.

c) The following data give the mandible lengths in millimeters for 10 male and ten female golden jackals in the collection of the British Museum. Using simulation, find the posterior density of the difference in mean mandible length between the sexes. Is there sufficient evidence to conclude that the males have a larger average?

Males

120 107 110 116 114 111 113 117 114 112

Females

110 111 107 108 110 105 107 106 111 111

3. **Comparing two proportions**

The following table gives the records of accidents in 1998 compiled by the Department of Highway Safety and Motor Vehicles in Florida.

	Injury	
Safety equipment in use	Fatal	Nonfatal
None	1601	162,527
Seat belt	510	412,368

Denote the number of accidents and fatalities when no safety equipment was in use by n_N and y_N, respectively. Similarly, let n_S and y_S denote the number of accidents and fatalities when a seat belt was in use. Assume that y_N and y_S are independent with y_N distributed binomial(n_N, p_N) and y_S distributed binomial(n_S, p_S). Assume a uniform prior is placed on the vector of probabilities (p_N, p_S).

a) Show that p_N and p_S have independent beta posterior distributions.

b) Use the function rbeta to simulate 1000 values from the joint posterior distribution of (p_N, p_S).

c) Using your sample, construct a histogram of the relative risk p_N/p_S. Find a 95% interval estimate of this relative risk.

d) Construct a histogram of the difference in risks $p_N - p_S$.

e) Compute the posterior probability that the difference in risks exceeds 0.

4. **Learning from rounded data**

It is a common problem for measurements to be observed in rounded form. Suppose we weigh an object five times and measure weights rounded to the nearest pound of 10, 1, 12, 11, 9. Assume the unrounded measurements are normally distributed with a noninformative prior distribution on the mean μ and variance σ^2.

a) Pretend that the observations are exact unrounded measurements. Simulate a sample of 1000 draws from the joint posterior distribution by using the algorithm described in Section 4.2.

b) Write down the correct posterior distributions for (μ, σ^2) treating the measurements as rounded.

c) By computing the correct posterior distribution on a grid of points (as in Section 4.4), simulate a sample from this distribution.

d) How do the incorrect and correct posterior distributions for μ compare? Answer this question by comparing posterior means and variances from the two simulated samples.

5. **Estimating the parameters of a Poisson/gamma density**
Suppose that $y_1, ..., y_n$ are a random sample from the Poisson/gamma density

$$f(y|a, b) = \frac{\Gamma(y + a)}{\Gamma(a)y!} \frac{b^a}{(b + 1)^{y+a}},$$

where $a > 0$ and $b > 0$. This density is an appropriate model for observed counts that show more dispersion than predicted under a Poisson model. Suppose that (a, b) are assigned the noninformative prior proportional to $1/(ab)$. If we transform to the real-valued parameters $\theta_1 = \log a$ and $\theta_2 = \log b$, the posterior density is proportional to

$$g(\theta_1, \theta_2|\text{data}) \propto \prod_{i=1}^{n} \frac{\Gamma(y_i + a)}{\Gamma(a)y_i!} \frac{b^a}{(b + 1)^{y_i+a}},$$

where $a = \exp\{\theta_1\}$ and $b = \exp\{\theta_2\}$. Use this framework to model data collected by Gilchrist (1984), in which a series of 33 insect traps were set across sand dunes and the numbers of different insects caught over a fixed time were recorded. The number of insects of the taxa *Staphylinoidea* caught in the traps are shown here.

$$2\ 5\ 0\ 2\ 3\ 1\ 3\ 4\ 3\ 0\ 3$$
$$2\ 1\ 1\ 0\ 6\ 0\ 0\ 3\ 0\ 1\ 1$$
$$5\ 0\ 1\ 2\ 0\ 0\ 2\ 1\ 1\ 1\ 0$$

By computing the posterior density on a grid, simulate 1000 draws from the joint posterior density of (θ_1, θ_2). From the simulated sample, find 90% interval estimates for the parameters a and b.

6. **Comparison of two Poisson rates** (from Antleman (1996))
A seller receives 800-number telephone orders from a first geographic area at a rate of λ_1 per week and from a second geographic area at a rate of λ_2 per week. Assume that incoming orders behave as if generated by a Poisson distribution; if the rate is λ, then the number of orders y in t weeks is distributed Poisson($t\lambda$). Suppose a series of newspaper ads are run in the first area for a period of four weeks, and sales for these four weeks are 260 units in area 1 and 165 units in area 2. The seller is interested in the effectiveness of these ads. One measure of this would be the probability that the sales rate in area 1 is greater than 1.5 times the sales rate in area 2:

$$P(\lambda_1 > 1.5\lambda_2).$$

Before the ads run, the seller has assessed a prior distribution for λ_1 to be gamma with parameters 144 and .417, and the prior for λ_2 to be gamma (100, .4).

a) Show that λ_1 and λ_2 have independent gamma posterior distributions.
b) Using the R function **rgamma**, simulate 1000 draws from the joint posterior distribution of (λ_1, λ_2).
c) Compute the posterior probability that the sales rate in area 1 is greater than 1.5 times the sales rate in area 2.

5

Introduction to Bayesian Computation

5.1 Introduction

In the previous two chapters, two types of strategies were used in the summarization of posterior distributions. If the sampling density has a familiar functional form, such as a member of an exponential family, and a conjugate prior is chosen for the parameter, then the posterior distribution often is expressible in terms of familiar probability distributions. In this case, we can simulate parameters directly by use of the R collection of random variate functions (such as `rnorm, rbeta` and `rgamma`), and we can summarize the posterior by computations on this simulated sample. A second type of computing strategy is what we called the "brute-force" method. In the case where the posterior distribution is not a familiar functional form, then one simply computes values of the posterior on a grid of points and then approximates the continuous posterior by a discrete posterior that is concentrated on the values of the grid. This brute-force method can be generally applied for one- and two-parameter problems such as those illustrated in Chapters 3 and 4.

In this chapter, we describe the Bayesian computational problem and introduce some of the more sophisticated computational methods that will be employed in later chapters. One general approach is based on the behavior of the posterior distribution about its mode. This gives a multivariate normal approximation to the posterior that serves as a good first approximation in the development of more exact methods. We then provide a general introduction to the use of simulation in computing summaries of the posterior distribution. When one can directly simulate samples from the posterior distribution, then the Monte Carlo algorithm gives an estimate and associated standard error for the posterior mean of any function of the parameters of interest. In the situation where the posterior distribution is not a standard functional form, rejection sampling with a suitable choice of proposal density provides an alternative method for producing draws from the posterior. Importance sampling and sampling importance resampling (SIR) algorithms are alternative general methods for computing integrals and simulating from a general posterior distribution. The SIR algorithm is especially useful when one wishes to investigate the sensitivity of a posterior distribution with respect to changes in the prior and likelihood functions.

5.2 Computing Integrals

The Bayesian recipe for inference is conceptually simple. If we observe data y from a sampling density $f(y|\theta)$, where θ is a vector of parameters and one assigns θ a prior $g(\theta)$, then the posterior density of θ is proportional to

$$g(\theta|y) \propto g(\theta)f(y|\theta).$$

The computational problem is to summarize this multivariate probability distribution to perform inference about functions of θ.

Many of the posterior summaries are expressible in terms of integrals. Suppose we are interested in the posterior mean of a function $h(\theta)$. This mean is expressible as a ratio of integrals

$$E(h(\theta)|y) = \frac{\int h(\theta)g(\theta)f(y|\theta)d\theta}{\int g(\theta)f(y|\theta)d\theta}.$$

If we are interested in the posterior probability that $h(\theta)$ falls in a set A, we wish to compute

$$P(h(\theta) \in A|y) = \frac{\int_{h(\theta) \in A} g(\theta)f(y|\theta)d\theta}{\int g(\theta)f(y|\theta)d\theta}.$$

Integrals are also involved when we are interested in obtaining marginal densities of parameters of interest. Suppose the parameter $\theta = (\theta_1, \theta_2)$, where θ_1 are the parameters of interest and θ_2 are so-called nuisance parameters. One obtains the marginal posterior density of θ_1 by integrating out the nuisance parameters from the joint posterior:

$$g(\theta_1|y) \propto \int g(\theta_1, \theta_2|y)d\theta_2.$$

In the common situation where one needs to evaluate these integrals numerically, there are a number of quadrature methods available. However, these quadrature methods have limited use for Bayesian integration problems. First, the choice of quadrature method depends on the location and shape of the posterior distribution. Second, for a typical quadrature method, the number of evaluations of the posterior density grows exponentially as a function of the number of components of θ. In this chapter, we focus on the use of computational methods for computing integrals that are applicable to high-dimensional Bayesian problems.

5.3 Setting Up a Problem on R

Before we describe some general summarization methods, we first describe setting up a Bayesian problem on R. Suppose one is able to write an explicit expression for the joint posterior density. In writing this expression, it is not necessary to include any normalizing constants that don't involve the parameters. Next, for the algorithms described in this book, it is helpful to reparameterize all parameters so that they are all real-valued. If one has a positive parameter such as a variance, then transform using a log function. If one has a proportion parameter p, then it can be transformed to the real line by the logit function $\text{logit}(p) = \log(p/(1-p))$.

After the posterior density has been expressed in terms of transformed parameters, the first step in summarizing this density is to write an R function defining the logarithm of the joint posterior density.

The general structure of this R function is

```
mylogposterior=function(theta,data)
{
[statements that compute the log density]
return(val)
}
```

To apply the functions described in this chapter, `theta` is assumed to be a matrix with n rows and k columns, where each row of `theta` corresponds to a value of the parameter vector $\theta = (\theta_1, ..., \theta_k)$. The input `data` is a vector of observed values or a list of data values and other model specifications such as the values of prior hyperparameters. The output vector `val` contains n values corresponding to the n values of the parameter vector θ.

One common situation is where one observes a random sample $y_1, ..., y_n$ from a sampling density $f(y|\theta)$ and one assigns θ the prior density. The logarithm of the posterior density of θ is given, up to an additive constant, by

$$\log g(\theta|y) = \log g(\theta) + \sum_{i=1}^{n} \log f(y_i|\theta).$$

When programming this function, it is important to note that the input is a matrix `theta` of parameter values. So it is necessary to use a loop to perform the summation when programming this function. Suppose we are sampling from a normal distribution with mean μ and standard deviation σ, the parameter vector $\theta = (\mu, \log \sigma)$ and we place an $N(10, 20)$ prior on μ and a flat prior on $\log \sigma$. The log posterior would have the form

$$\log g(\theta|y) = \log \phi(\mu; 10, 20) + \sum_{i=1}^{n} \log \phi(y_i; \mu, \sigma),$$

where $\phi(y; \mu, \sigma)$ is the normal density with mean μ and standard deviation σ. If `data` is the vector of observations $y_1, ..., y_n$, then the function defining the log posterior would in this case would be written as follows.

```
mylogposterior=function(theta,data)
{
n=length(data)
mu=theta[,1];  sigma=exp(theta[,2])
val=0*mu
for (i in 1:n)
{
val=val+dnorm(data[i],mean=mu,sd=sigma,log=TRUE)
}
val=val+dnorm(mu, mean=10, sd=20,log=TRUE)
return(val)
}
```

We use the `log = TRUE` option in `dnorm` to compute the logarithm of the density. Note the use of the small trick `val=0*mu`; this is a simple way of creating a zero column vector of the same size as the vector `mu`.

5.4 A Beta-Binomial Model for Overdispersion

Tsutakawa et al (1985) describe the problem of simultaneously estimating the rates of death from stomach cancer for males at risk in the age bracket 45–64 for the largest cities in Missouri. Table 5.1 displays the mortality rates for 20 of these cities, where a cell contains the number n_j at risk and the number of cancer deaths y_j for a given city.

Table 5.1. Cancer mortality data. Each ordered pair represents the number of cancer deaths y_j and the number at risk n_j for an individual city in Missouri.

(0, 1083)	(0, 855)	(2, 3461)	(0, 657)	(1, 1208)	(1, 1025)
(0, 527)	(2, 1668)	(1, 583)	(3, 582)	(0, 917)	(1, 857)
(1, 680)	(1, 917)	(54, 53637)	(0, 874)	(0, 395)	(1, 581)
(3, 588)	(0, 383)				

A first modeling attempt might assume that the $\{y_j\}$ represent independent binomial samples with sample sizes $\{n_j\}$ and common probability of death p. But it can be shown that these data are overdispersed in the sense that the counts $\{y_j\}$ display more variation that would be predicted under a binomial model with a constant probability p. A better fitting model assumes that y_j is distributed from a beta-binomial model with mean η and precision K:

$$f(y_j|\eta, K) = \binom{n_j}{y_j} \frac{B(K\eta + y_j, K(1 - \eta) + n_j - y_j)}{B(K\eta, K(1 - \eta))}.$$

Suppose we assign the parameters the vague prior proportional to

$$g(\eta, K) \propto \frac{1}{\eta(1-\eta)} \frac{1}{(1+K)^2}.$$

Then the posterior density of (η, K) is given, up to a proportionality constant, by

$$g(\eta, K) \propto \frac{1}{\eta(1-\eta)} \frac{1}{(1+K)^2} \prod_{j=1}^{20} \frac{B(K\eta + y_j, K(1-\eta) + n_j - y_j)}{B(K\eta, K(1-\eta))},$$

where $0 < \eta < 1$ and $K > 0$.

We write a short function betabinexch0 to compute the logarithm of the posterior density. The inputs to the function are theta, a matrix where the values of η and K are respectively in the first and second columns, and data, a matrix with columns the vector of counts $\{y_j\}$ and the vector of sample sizes $\{n_j\}$.

```
betabinexch0=function(theta,data)
{
eta=theta[,1]
K=theta[,2]
y=data[,1]; n=data[,2]
N=length(y)
val=0*K;
for (i in 1:N)
    val=val+lbeta(K*eta+y[i],K*(1-eta)+n[i]-y[i])
val=val-N*lbeta(K*eta,K*(1-eta))
val=val-2*log(1+K)-log(eta)-log(1-eta)
return(val)
}
```

We read in the dataset cancermortality and use the function mycontour together with the log density function betabinexch0 to display a contour plot of the posterior density of (η, K) (See Fig. 5.1).

```
> data(cancermortality)
> mycontour(betabinexch0,c(.0001,.003,1,20000),cancermortality)
```

Note the strong skewness in the density, especially toward large values of the precision parameter K. This right skewness is a common characteristic of the likelihood function of a precision or variance parameter. Following the general guidance in Section 5.3, suppose we transform each parameter to the real line by the reexpressions

$$\theta_1 = \text{logit}(\eta) = \log\left(\frac{\eta}{1-\eta}\right), \quad \theta_2 = \log(K).$$

The log posterior density of the transformed parameters is programmed in the function betabinexch. Note the change in the next-to-last line of the function that accounts for the Jacobian term in the transformation.

```
betabinexch=function(theta,data)
{
theta1=theta[,1]
theta2=theta[,2]
eta=exp(theta1)/(1+exp(theta1))
K=exp(theta2)
y=data[,1]; n=data[,2]
N=length(y);
val=0*K;
for (i in 1:N)
    val=val+lbeta(K*eta+y[i],K*(1-eta)+n[i]-y[i])
val=val-N*lbeta(K*eta,K*(1-eta))
val=val+theta2-2*log(1+exp(theta2))
return(val)
}
```

Fig. 5.2 displays a contour plot of the posterior of (θ_1, θ_2) using the mycontour function. Although the density has an unusual shape, the strong skewness has been reduced and the distribution is more amenable to the computational methods described in this and the following chapters.

5.5 Approximations Based on Posterior Modes

One method of summarizing a multivariate posterior distribution is based on the behavior of the density about its mode. Let θ be a vector-valued parameter with prior density $g(\theta)$. If we observe data y with sampling density $f(y|\theta)$, then consider the logarithm of the joint density of θ and y

$$h(\theta, y) = \log(g(\theta)f(y|\theta)).$$

In the following, we write this log density as $h(\theta)$ since after the data are observed θ is the only random quantity. Denoting the posterior mode of θ by $\hat{\theta}$, we expand the log density in a second-order Taylor series about $\hat{\theta}$. This gives the approximation

$$h(\theta) \approx h(\hat{\theta}) + (\theta - \hat{\theta})'h''(\hat{\theta})(\theta - \hat{\theta})/2,$$

where $h''(\hat{\theta})$ is the Hessian of the log density evaluated at the mode. By this expansion, the posterior density is approximated by a multivariate normal density with mean $\hat{\theta}$ and variance-covariance matrix

$$V = (-h''(\hat{\theta}))^{-1}.$$

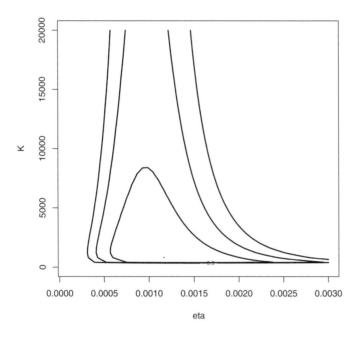

Fig. 5.1. Contour plot of parameters η and K in the beta-binomial model problem.

In addition, this approximation allows one to analytically integrate out θ from the joint density and obtain the following approximation to the prior predictive density:

$$f(y) \approx (2\pi)^{d/2} g(\hat{\theta}) f(y|\hat{\theta})| - h''(\hat{\theta})|^{1/2},$$

where d is the dimension of θ.

To apply this approximation, one needs to find the mode of the posterior density of θ. A good general-purpose optimization algorithm for finding this mode is provided by Newton's method. Suppose one has a guess at the posterior mode θ^0. If θ^{t-1} is the estimate at the mode at the $t-1$ iteration of the algorithm, then the next iterate is given by

$$\theta^t = \theta^{t-1} - [h''(\theta^{t-1})]^{-1} h'(\theta^{t-1}).$$

One continues these iterations until convergence.

After one writes an R function to evaluate the log posterior density, the R function `laplace` in the LearnBayes package finds the joint posterior mode by several iterations of Newton's method. The inputs to `laplace` are the function defining the joint posterior, an intelligent guess at the posterior mode,

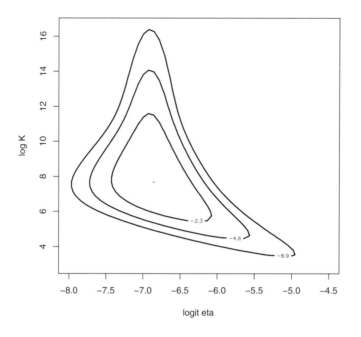

Fig. 5.2. Contour plot of transformed parameters $\mathrm{logit}(\eta)$ and $\log K$ in the beta-binomial model problem.

the number of iterations of this algorithm, and data and parameters used in the definition of the log posterior. The choice of "intelligent guess" can be important since Newton's algorithm may fail to converge with a poor choice of starting value.

Suppose that a suitable starting value is used and `laplace` is successful in finding the posterior mode. The output of `laplace` is a list with three components. The component `mode` gives the value of the posterior mode $\hat{\theta}$, the component `var` is the associated variance-covariance matrix V, and the component `int` is the approximation to the logarithm of the prior predictive density.

5.6 The Example

We illustrate the use of the function `laplace` for our beta-binomial modeling example. Based on our contour plot, we start Newton's method with the initial guess $(\mathrm{logit}(\eta), \log K) = (-7, 6)$ and perform 10 Newton steps.

```
> fit=laplace(betabinexch,array(c(-7,6),c(1,2)),10,cancermortality)
> fit
```

`$mode`

```
          [,1]     [,2]
[1,] -6.818793 7.57451
```

`$var`

```
          [,1]          [,2]
[1,]   0.07903107 -0.1490403
[2,] -0.14904028  1.3490592
```

`$int`

```
[1] -570.7744
```

We find the posterior mode to be $(-6.82, 7.57)$. Also this gives the approximation that $(\text{logit}(\eta), \log K)$ is approximately bivariate normal with mean vector fit$mode and variance-covariance matrix fit$var. By use of the mycontour function with the log bivariate normal function lbinorm, Fig. 5.3 displays the contours of the approximate normal density. Comparing Fig. 5.2 and Fig.5.3, we see significant differences between the exact and approximate normal posteriors.

```
> npar=list(m=fit$mode,v=fit$var)
> mycontour(lbinorm,c(-8,-4.5,3,16.5),npar)
> title(xlab="logit eta", ylab="log K")
```

One advantage of this algorithm is that one obtains quick summaries of the parameters by use of the multivariate normal approximation. By use of the diagonal elements of the variance-covariance matrix, one can construct approximate probability intervals for $\text{logit}(\eta)$ and $\log K$. For example, the following code constructs 90% probability intervals for the parameters:

```
> se=diag(fit$var)
> fit$mode-1.645*se
```

```
          [,1]     [,2]
[1,] -6.948801 5.35523
```

```
> fit$mode+1.645*se
```

```
          [,1]     [,2]
[1,] -6.688786 9.79379
```

So a 90% interval estimate for $\text{logit}(\eta)$ is $(-6.95, -6.69)$, and a 90% interval estimate for $\log K$ is $(5.36, 9.79)$.

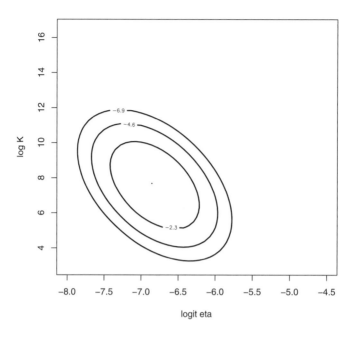

Fig. 5.3. Contour plot of normal approximation of $\text{logit}(\eta)$ and $\log K$ in the beta-binomial model problem.

5.7 Monte Carlo Method for Computing Integrals

A second general approach for summarizing a posterior distribution is based on simulation. Suppose that θ has a posterior density $g(\theta|y)$ and we are interested in learning about a particular function of the parameters $h(\theta)$. The posterior mean of $h(\theta)$ is given by

$$E(h(\theta)|y) = \int h(\theta)g(\theta|y)d\theta.$$

Suppose we are able to simulate an independent sample $\theta^1, ..., \theta^m$ from the posterior density. Then the Monte Carlo estimate at the posterior mean is given by the sample mean

$$\bar{h} = \frac{\sum_{j=1}^{m} h(\theta^j)}{m}.$$

The associated simulation standard error of this estimate is estimated by

$$se_{\bar{h}} = \sqrt{\frac{\sum_{j=1}^{m}(h(\theta^j) - \bar{h})^2}{(m-1)m}}.$$

The Monte Carlo approach is an effective method for summarizing a posterior distribution when simulated samples are available from the exact posterior distribution. For a simple illustration of the Monte Carlo method, return to Section 2.4 where we were interested in the proportion of heavy sleepers p at a college. With the use of a beta prior, the posterior distribution for p was beta(14.4, 23.4). Suppose we are interested in the posterior mean of p^2. (This is the predictive probability that two students in a future sample will be heavy sleepers.) We simulate 1000 draws from the beta posterior distribution. If $\{p^j\}$ represent the simulated sample, the Monte Carlo estimate at this posterior mean will be the mean of the $\{(p^j)^2\}$, and the simulated standard error is the standard deviation of the $\{(p^j)^2\}$ divided by the square root of the simulation sample size.

```
> p=rbeta(1000, 14.4, 23.4)
> est=mean(p^2)
> se=sd(p^2)/sqrt(1000)
> c(est,se)
```

```
[1] 0.152099714 0.001925061
```

The Monte Carlo estimate at $E(p^2|\text{data})$ is 0.152 with an associated simulation standard error of 0.002.

5.8 Rejection Sampling

In the examples of Chapter 2, 3, and 4, we were able to produce simulated samples directly from the posterior distribution since the distributions were familiar functional forms. Then we would be able to obtain Monte Carlo estimates at the posterior mean of any function of the parameters of interest. But in many situations such as the beta-binomial example of this chapter, the posterior does not have a familiar form and we need to use an alternative algorithm for producing a simulated sample.

A general-purpose algorithm for simulating random draws from a given probability distribution is rejection sampling. In this setting, suppose we wish to produce an independent sample from a posterior density $g(\theta|y)$ where the normalizing constant may not be known. The first step in rejection sampling is to find another probability density $p(\theta)$ such that

- It is easy to simulate draws from p.
- The density p resembles the posterior density of interest g in terms of location and spread.
- For all θ and a constant c, $g(\theta|y) \leq cp(\theta)$.

Suppose we are able to find a density p with these properties. Then one obtains draws from g by the following accept/reject algorithm:

1. Simulate independently θ from p and a uniform random variable U on the unit interval.
2. If $U \leq g(\theta|y)/(cp(\theta))$, then accept θ as a draw from the density g, otherwise reject θ.
3. Continue steps 1 and 2 of the algorithm until one has collected a sufficient number of "accepted" θ.

Rejection sampling is one of the most useful methods for simulating draws from a variety of distributions and standard methods for simulating from standard probability distributions such as normal, gamma, and beta are typically based on rejection algorithms. The main task in designing a rejection sampling algorithm is finding a suitable proposal density p and constant value c. At step 2 of the algorithm, the probability of accepting a candidate draw is given by $g(\theta|y)/(cp(\theta))$. One can monitor the algorithm by computing the proportion of draws of p that are accepted; an efficient rejection sampling algorithm has a high acceptance rate.

We consider the use of rejection sampling to simulate draws of $\theta = (\mathrm{logit}(\eta), \log K)$ in the beta-binomial example. We wish to find a proposal density of a simple functional form that, when multiplied by an appropriate constant, covers the posterior density of interest. One choice for p would be a bivariate normal density with mean and variance given as outputs of the function `laplace`. Although this density does resemble the posterior density, the normal density has relatively sharp tails and likely the ratio $g(\theta|y)/p(\theta)$ would not be bounded. A better choice for a covering density is a multivariate t with mean and scale matrix chosen to match the posterior density and a small number of degrees of freedom. The small number of degrees of freedom gives the density heavy tails and one is more likely to find bounds for the ratio $g(\theta|y)/p(\theta)$.

In our earlier work, we found approximations to the posterior mean and variance-covariance matrix of $\theta = (\mathrm{logit}(\eta), \log K)$ based on the Laplace method. If the output variable of `laplace` is `fit`, then `fit$mode` is the posterior mode and `fit$var` the associated variance-covariance matrix. Suppose we decide to use a multivariate t density with location `fit$mode`, scale matrix `2 fit$var`, and four degrees of freedom. These choices are made to mimic the posterior density and ensure that the ratio $g(\theta|y)/p(\theta)$ is bounded from above.

To set up, we need to find the value of the bounding constant. We want to find the constant c such that

$$g(\theta|y) \leq cp(\theta) \text{ for all } \theta.$$

Equivalently, since g is programmed on the log scale, we want to find the constant $d = \log c$ such that

$$\log g(\theta|y) - \log p(\theta) \leq d \text{ for all } \theta.$$

Basically we wish to maximize the function $\log g(\theta|y) - \log p(\theta)$ over all θ. A convenient way to perform this maximization is by use of the `laplace` function. We write a new function `betabinT` that computes values of this difference function. There are two inputs, the parameter `theta` and a list `datapar` with components `data`, the data matrix, and `par`, a list with the parameters of the t proposal density (mean, scale matrix, and degrees of freedom)

```
betabinT=function(theta,datapar)
{
data=datapar$data
tpar=datapar$par
d=betabinexch(theta,data)-dmt(theta,mean=c(tpar$m),
  S=tpar$var,df=tpar$df,log=TRUE)
return(d)
}
```

For our problem, we define the parameters of the t proposal density and the list `datapar`:

```
> tpar=list(m=fit$mode,var=2*fit$var,df=4)
> datapar=list(data=cancermortality,par=tpar)
```

We run the function `laplace` with this new function and use of an "intelligent" starting value.

```
> start=array(c(-6.9,12.4),c(1,2))
> fit1=laplace(betabinT,start,10,datapar)
> fit1$mode
```

```
        [,1]      [,2]
[1,] -6.889 12.42736
```

We find the maximum value d occurs at the value $\theta = (-6.889, 12.42736)$. We note that this θ value is not at the extreme portion of the space of simulated draws that indicates that we indeed have found an approximate maximum. The value of d is found by evaluating the function at the modal value.

```
> betabinT(fit1$mode,datapar)
```

```
[1] -569.2813
```

We implement rejection sampling using the function `rejectsampling`. The inputs are the function defining the log posterior, the parameters of the t covering density, the value of d, the number of candidate values simulated, and the data for the log posterior function. In this function, we simulate a vector of θ from the proposal density, compute the values of $\log g$ and $\log f$ on these simulated draws, compute the acceptance probabilities, and return only the simulated values of θ where the uniform draws are smaller than the acceptance probabilities.

```
rejectsampling=function(logf,tpar,dmax,n,data)
{
    theta=rmt(n,mean=c(tpar$m),S=tpar$var,df=tpar$df)
    lf=logf(theta,data)
    lg=dmt(theta,mean=c(tpar$m),S=tpar$var,df=tpar$df,log=TRUE)
    prob=exp(lf-lg-dmax)
    return(theta[runif(n)<prob,])
}
```

We run the function rejectsampling using the constant value of d found earlier and simulate 10,000 draws from the proposal density. We see that the output value theta has only 2406 rows, so the acceptance rate of this algorithm is $2406/10{,}000 = .24$. This is a relatively inefficient algorithm since it has a small acceptance rate, but the proposal density was found without too much effort.

```
> theta=rejectsampling(betabinexch,tpar,-569.2813,10000,
cancermortality)
> dim(theta)
```

```
[1] 2406    2
```

We plot the simulated draws from rejection sampling on the contour plot of the log posterior density in Fig. 5.4. As expected, most of the draws fall within the inner contour of the exact density.

```
> mycontour(betabinexch,c(-8,-4.5,3,16.5),cancermortality)
> points(theta[,1],theta[,2])
```

5.9 Importance Sampling

Let us return to the basic problem of computing an integral in Bayesian inference. In many situations, the normalizing constant of the posterior density $g(\theta|y)$ will be unknown. So the posterior mean of the function $h(\theta)$ will be given by the ratio of integrals

$$E(h(\theta)|y) = \frac{\int h(\theta)g(\theta|y)d\theta}{\int g(\theta|y)d\theta}.$$

If we were able to simulate a sample $\{\theta^j\}$ directly from the posterior density g, then one could approximate this expectation by a Monte Carlo estimate. In the case where we are not able to generate a sample directly from g, suppose instead that we can construct a probability density p that we can simulate and that approximates the posterior density g. We rewrite the posterior mean as

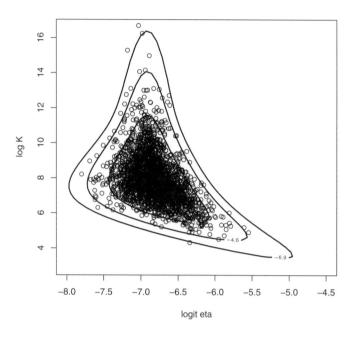

Fig. 5.4. Contour plot of logit(η) and $\log K$ in the beta-binomial model problem together with simulated draws from the rejection algorithm.

$$E(h(\theta)|y) = \frac{\int h(\theta)\frac{g(\theta|y)}{p(\theta)}p(\theta)d\theta}{\int \frac{g(\theta|y)}{p(\theta)}p(\theta)d\theta}$$

$$= \frac{\int h(\theta)w(\theta)p(\theta)d\theta}{\int w(\theta)p(\theta)d\theta},$$

where $w(\theta) = g(\theta|y)/p(\theta)$ is the weight function. If $\theta^1, ..., \theta^m$ are a simulated sample from the approximation density p, then the importance sampling estimate at the posterior mean is

$$\bar{h}_{IS} = \frac{\sum_{j=1}^m h(\theta^j)w(\theta^j)}{\sum_{j=1}^m w(\theta^j)}.$$

This is called an *importance sampling estimate* because we are sampling values of θ that are important in computing the integrals in the numerator and denominator. The simulation standard error of an importance sampling estimate is estimated by

$$se_{\bar{h}_{IS}} = \sqrt{\frac{\sum_{j=1}^m ((h(\theta^j) - \bar{h}_{IS})w(\theta^j))^2}{\sum_{j=1}^m w(\theta^j)}}.$$

As in rejection sampling, the main issue in designing a good importance sampling estimate is finding a suitable sampling density p. This density should be of a familiar functional form so simulated draws are available. The density should mimic the posterior density g and have relatively flat tails so that the weight function $w(\theta)$ is bounded from above. One can monitor the choice of p by inspecting the values of the simulated weights $w(\theta^j)$. If there are not any unusually large weights, then it is likely that the weight function is bounded and the importance sampler is providing a suitable estimate.

To illustrate importance sampling, let us return to our beta-binomial example and consider the problem of estimating the posterior mean of $\log K$. For a posterior density of real-valued parameters, a convenient choice of sampler p is a multivariate t density. The R function `impsampling` will implement importance sampling when p is a t density. There are five inputs to this function: `logf` is the function defining the logarithm of the posterior, `tpar` is a list of parameter values of the t density, `h` is a function defining the function $h(\theta)$ of interest, `n` is the size of the simulated sample, and `data` is the vector or list used in the definition of `logf`. In the function `impsampling`, the functions `rmt` and `dmt` from the `mnormt` library are used to simulate and compute values of the t density. Note that the value `md` is the maximum value of the difference of logs of the posterior and proposing density – this value is used in the computation of the weights to prevent possible overflow. The output of `impsampling` is a list with four components: `est` is the importance sampling estimate, `se` is the corresponding simulation standard error, `theta` is a matrix of simulated draws from the proposing density p, and `wt` is a vector of the corresponding weights.

```
impsampling=function(logf,tpar,h,n,data)
{
theta=rmt(n,mean=c(tpar$m),S=tpar$var,df=tpar$df)
lf=logf(theta,data)
lp=dmt(theta,mean=c(tpar$m),S=tpar$var,df=tpar$df,log=TRUE)
md=max(lf-lp)
wt=exp(lf-lp-md)
est=sum(wt*h(theta))/sum(wt)
SEest=sqrt(sum((h(theta)-est)^2*wt^2))/sum(wt)
return(list(est=est,se=SEest,theta=theta,wt=wt))
}
```

For this example, the choice of proposal density used in the development of a rejection algorithm seems to be a good choice for importance sampling. We choose a t density where the location is the posterior mode (found from `laplace`), the scale matrix is twice the estimated variance-covariance matrix, and the number of degrees of freedom is four. This choice for p will resemble the posterior density and have flat tails that we hope will result in bounded weights. We define a short function `myfunc` to compute the function h; since we are interested in the posterior mean of $\log K$ we define the function to be the second column of the matrix θ. We are now ready to run impsampling.

```
> tpar=list(m=fit$mode,var=2*fit$var,df=4)
> myfunc=function(theta)
+    return(theta[,2])
> s=impsampling(betabinexch,tpar,myfunc,10000,cancermortality)
> cbind(s$est,s$se)

          [,1]          [,2]
[1,] 7.957802 0.01967276
```

We see from the output that the importance sampling estimate of the mean of $\log K$ is 7.958 with an associated standard error of 0.020. To check if the weight function is bounded, we compute a histogram of the simulated weights (not shown here) and note that there are no extreme weights.

5.10 Sampling Importance Resampling

In rejection sampling, we simulated draws from a proposal density p and accepted a subset of these values to be distributed according to the posterior density of interest $g(\theta|y)$. There is an alternative method of obtaining a simulated sample from the posterior density g motivated by the importance sampling algorithm.

As before, we simulate m draws from the proposal density p denoted by $\theta^1, ..., \theta^m$ and compute the weights $\{w(\theta^j) = g(\theta^j|y)/p(\theta^j)\}$. Convert the weights to probabilities by the formula

$$p^j = \frac{w(\theta^j)}{\sum_{j=1}^m w(\theta^j)}.$$

Suppose we take a new sample $\theta^{*1}, ..., \theta^{*m}$ from the discrete distribution over $\theta^1, ..., \theta^m$ with respective probabilities $p^1, ..., p^m$. Then the $\{\theta^{*j}\}$ will be approximately distributed according to the posterior distribution g. This method, called sampling importance sampling or SIR for short, is a weighted bootstrap procedure where we sample with replacement from the sample $\{\theta^j\}$ with unequal sampling probabilities.

This sampling algorithm is straightforward to implement in R using the sample command. Suppose we wish to obtain a simulated sample of size n. As in importance sampling, we first simulate from the proposal density which in this situation is a multivariate t distribution, and then compute the importance sampling weights stored in the vector wt.

```
theta = rmt(n, mean = c(tpar$m), S = tpar$var, df = tpar$df)
lf = logf(theta, data)
```

```
lp = dmt(theta, mean = c(tpar$m), S = tpar$var, df = tpar$df,
         log = TRUE)
md = max(lf - lp)
wt = exp(lf - lp - md)
```

To implement the SIR algorithm, we first convert the weights to probabilities and store them in the vector `probs`. Next we use `sample` to take a sample with replacement from the indices 1, ..., n, where the sampling probabilities are contained in the vector `probs`; the simulated indices are stored in the vector `indices`.

```
probs=wt/sum(wt)
indices=sample(1:n,size=n,prob=probs,replace=TRUE)
```

Finally, we use the random indices in `indices` to select the rows of `theta` and assign to the matrix `theta.s`. The matrix `theta.s` contain the simulated draws from the posterior.

```
theta.s=theta[indices,]
```

The function `sir` implements this algorithm for a multivariate t proposal density. The inputs to this function are the function defining the log posterior `logf`, the list `tpar` of parameters of the multivariate proposal density, the number `n` of simulated draws, and the `data` used in the log posterior function. The output is a matrix of simulated draws from the posterior. In the beta-binomial modeling example, we implement the SIR algorithm by the command

```
> theta.s=sir(betabinexch,tpar,10000,cancermortality)
```

We have illustrated the use of the SIR algorithm in converting simulated draws from a proposal density to draws from the posterior density. But this algorithm can be used to convert simulated draws from one probability density to a second probability density. To show the power of this method, suppose we wish to perform a Bayesian sensitivity analysis with respect to the individual observations in the dataset. Suppose we focus on posterior inference about the log precision parameter $\log K$ and question how the inference would change if we removed individual observations from the likelihood. Let $g(\theta|y)$ denote the posterior density from the full dataset and $g(\theta|y_{(-i)})$ denote the posterior density with the ith observation removed. Let $\{\theta^j\}$ represent a simulated sample from the full dataset. We can obtain a simulated sample from $g(\theta|y_{(-i)})$ by resampling from $\{\theta^j\}$, where the sampling probabilities are proportional to the weights

$$
w(\theta) = \frac{g(\theta|y_{(-i)})}{g(\theta|y)}
$$
$$
= \frac{1}{f(y_i|\theta)}
$$
$$
= \frac{B(K\eta, K(1-\eta))}{B(K\eta + y_i, K(1-\eta) + n_i - y_i)}.
$$

Suppose that the inference of interest is a 90% probability interval for the log precision $\log K$. The R code for this resampling for the "leave observation i out" follows. One first computes the sampling weights and the sampling probabilities. Then the `sample` command is used to do the resampling from `theta` and the simulated draws from the "leave one out" posterior are stored in the variable `theta.s`. We summarize the simulated values of $\log K$ by the 5th, 50th, and 95th quantiles.

```
weight=exp(lbeta(K*eta,K*(1-eta))-
    lbeta(K*eta+y[i],K*(1-eta)+n[i]-y[i]))
probs=weight/sum(weight)
indices=sample(1:m,size=m,prob=probs,replace=TRUE)
theta.s=theta[indices,]
summary.obs[i,]=quantile(theta.s[,2],c(.05,.5,.95))
```

The function `bayes.influence` computes probability intervals for $\log K$ for the complete dataset and "leave one out" datasets using the SIR algorithm. We assume one already has simulated a sample of values from the complete data posterior and the draws are stored in the matrix variable `theta.s`. The inputs to `bayes.influence` are `theta.s` and the dataset `data`. In this case, suppose we have just implemented the SIR algorithm and the posterior draws are stored in the matrix `theta.s`. Then the form of the function would be

```
> S=bayes.influence(theta.s,cancermortality)
```

The output of this function is a list S; S$summary is a vector containing the 5th, 50th, and 95th percentiles and S$summary.obs is a matrix where the ith row gives the percentiles for the posterior with the ith observation removed.

Fig. 5.5 is a graphical display of the sensitivity of the posterior inference about $\log K$ with respect to the individual observations. The bold line shows the posterior median and 90% probability interval for the complete dataset and the remaining lines show the inference with each possible observation removed. Note that if observation number 15 is removed $((y_i, n_i) = (54, 53637))$, then the location of $\log K$ is shifted toward smaller values. Also if either observation 10 or observation 19 is removed, $\log K$ is shifted toward larger values. These two observations are notable since each city experienced three deaths and had relatively high mortality rates.

```
> plot(c(0,0,0),S$summary,type="b",lwd=3,xlim=c(-1,21),
+   ylim=c(5,11), xlab="Observation removed",ylab="log K")
> for (i in 1:20)
+   lines(c(i,i,i),S$summary.obs[i,],type="b")
```

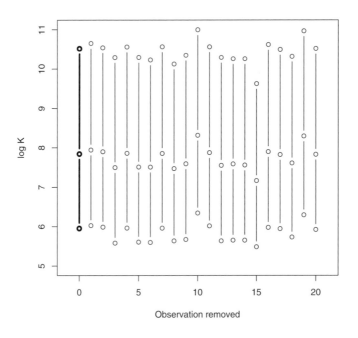

Fig. 5.5. Ninety percent interval estimates for log K for full dataset (thick line) and interval estimates for datasets with each individual observation removed.

5.11 Further Reading

Rejection sampling is a general method used in simulating probability distributions; discussion of rejection sampling for statistical problems is described in Givens and Hoeting (2005), Monahan (2001), and Robert and Casella (2004). Tanner (1996) introduces normal approximations to posterior distributions in chapter 2 and Monte Carlo methods in chapter 3. Robert and Casella (2004) in chapter 3 describe different aspects of Monte Carlo integration. Smith and Gelfand (1992) introduce the use of rejecting sampling and the SIR algorithm in simulating from the posterior distribution.

5.12 Summary of R Functions

`bayes.influence` – computes probability intervals for the log precision parameter K in a beta-binomial model for all "leave one out" models using sampling importance resampling
Usage: `bayes.influence(theta,data)`

Arguments: theta, matrix of simulated draws from the posterior of (logit eta, log K) for a beta-binomial model; data, matrix with columns of counts and sample sizes

Value: summary, vector of 5th, 50th and 95th percentiles of log K for posterior of complete sample; summary.obs, matrix where the *i*th row contains the 5th, 50th and 95th percentiles of log K for posterior when the ith observation is removed

betabinexch0 – computes the logarithm of the posterior for the parameters (mean and precision) in a beta/binomial model

Usage: betabinexch0(theta,data)

Arguments: theta, matrix of parameter values where each row represents a value of (eta, K); data, matrix with columns of counts and sample sizes

Value: vector of values of the log posterior where each value corresponds to each row in the parameters in theta

betabinexch – computes the logarithm of the posterior for the parameters (logit mean and log precision) in a beta/binomial model

Usage: betabinexch(theta,data)

Arguments: theta, matrix of parameter values where each row represents a value of (logit eta, log K); data, matrix with columns of counts and sample sizes

Value: vector of values of the log posterior where each value corresponds to each row in the parameters in theta

impsampling – implements importance sampling to compute the posterior mean of a function using a multivariate t proposal density

Usage: impsampling(logf,tpar,h,n,data)

Arguments: logf, function defining the log density; tpar, list of parameters of a multivariate t proposal density including the mean m, the scale matrix var, and the degrees of freedom df; h, function that defines h(theta); n, number of simulated draws from the proposal density; data, data and or parameters used in the function logf

Value: est, estimate at the posterior mean; se, simulation standard error of the estimate; theta, matrix of simulated draws from proposal density; wt, vector of importance sampling weights

laplace – for a general posterior density, computes the posterior mode, the associated variance-covariance matrix, and an estimate at the logarithm at the normalizing constant

Usage: laplace(logpost,mode,iter,par)

Arguments: logpost, function that defines the logarithm of the posterior density; mode, vector that is a guess at the posterior mode; iter, number of iterations of Newton-Raphson algorithm; par, vector or list of parameters associated with the function logpost

Value: `mode`, current estimate at the posterior mode; `var`, current estimate at the associated variance-covariance matrix; `int`, estimate at the logarithm of the normalizing constant

`rejectsampling` – implements a rejection sampling algorithm for a probability density using a multivariate t proposal density
Usage: `rejectsampling(logf,tpar,dmax,n,data)`
Arguments: `logf`, function that defines the logarithm of the density of interest; `tpar`, list of parameters of a multivariate t proposal density including the mean `m`, the scale matrix `var`, and the degrees of freedom `df`; `dmax`, logarithm of the rejection sampling constant; `n`, number of simulated draws from the proposal density; `data`, data and or parameters used in the function logf
Value: matrix of simulated draws from density of interest

`sir` – implements the sampling importance resampling algorithm for a multivariate t proposal density
Usage: `sir(logf,tpar,n,data)`
Arguments: `logf`, function defining logarithm of density of interest; `tpar`, list of parameters of a multivariate t proposal density including the mean `m`, the scale matrix `var`, and the degrees of freedom `df`; `n`, number of simulated draws from the posterior; `data`, data and parameters used in the function logf
Value: matrix of simulated draws from the posterior where each row corresponds to a single draw

5.13 Exercises

1. **Estimating a log-odds with a normal prior**
   Suppose y has a binomial distribution with parameters n and p, and we are interested in the log-odds value $\theta = \log(p/(1-p))$. Our prior for θ is that $\theta \sim N(\mu, \sigma)$. It follows that the posterior density of θ is given, up to a proportionality constant, by

$$g(\theta|y) \propto \frac{\exp(y\theta)}{(1+\exp(\theta))^n} \exp\left[\frac{-(\theta-\mu)^2}{2\sigma^2}\right].$$

   More concretely, suppose we are interested in learning about the probability a special coin lands heads when tossed. A priori we believe that the coin is fair, so we assign θ an $N(0, .25)$ prior. We toss the coin $n = 5$ times and obtain $y = 5$ heads.
   a) Using a normal approximation to the posterior density, compute the probability that the coin is biased toward heads (i.e., that θ is positive).
   b) Using the prior density as a proposal density, design a rejection algorithm for sampling from the posterior distribution. Using simulated draws from your algorithm, approximate the probability that the coin is biased toward heads.

c) Using the prior density as a proposal density, simulate values from the posterior distribution using the SIR algorithm. Approximate the probability the coin is biased toward heads.

2. **Genetic linkage model from** Rao (2002)

Suppose 197 animals are distributed into four categories with the following frequencies:

Category	1	2	3	4
Frequency	125	18	20	34

Assume that the probabilities of the four categories are given by the vector

$$\left(\frac{1}{2} + \frac{\theta}{4}, \frac{1}{4}(1 - \theta), \frac{1}{4}(1 - \theta), \frac{\theta}{4}\right),$$

where θ is an unknown parameter between 0 and 1. If θ is assigned a uniform prior, then the posterior density of θ is given by

$$g(\theta|\text{data}) \propto \left(\frac{1}{2} + \frac{\theta}{4}\right)^{125} \left(\frac{1}{4}(1 - \theta)\right)^{18} \left(\frac{1}{4}(1 - \theta)\right)^{20} \left(\frac{\theta}{4}\right)^{34},$$

where $0 < \theta < 1$. If θ is transformed to the real-valued logit $\eta = \log(\theta/(1 - \theta))$, then the posterior density of η can be written as

$$f(\eta|\text{data}) \propto \left(2 + \frac{e^\eta}{1 + e^\eta}\right)^{125} \frac{1}{(1 + e^\eta)^{39}} \left(\frac{e^\eta}{1 + e^\eta}\right)^{35}, \quad -\infty < \eta < \infty.$$

a) Use a normal approximation to find a 95% probability interval for η. Transform this interval to obtain a 95% probability interval for the original parameter of interest θ.

b) Design a rejection sampling algorithm for simulating from the posterior density of η. Use a t proposal density using a small number of degrees of freedom and mean and scale parameters given by the normal approximation.

3. **Estimation for the two-parameter exponential distribution**

Martz and Waller (1982) describe the analysis of a "type I/time-truncated" life testing experiment. Fifteen reciprocating pumps were tested for a pre-specified time; any failures among the pumps were replaced. One assumes that the failure times follow the two-parameter exponential distribution

$$f(y|\beta, \mu) = \frac{1}{\beta} e^{-(y-\mu)/\beta}, \quad y \geq \mu.$$

Suppose one places a uniform prior on (μ, β). Then Martz and Waller show that the posterior density is given by

$$g(\beta, \mu|\text{data}) \propto \frac{1}{\beta^s} \exp\{-(t - n\mu)/\beta\}, \quad \mu \leq t_1,$$

where n is the number of items placed on test, t is the total time on test, t_1 is the smallest failure time, and s is the observed number of failures in a sample of size n. In the example, data were reported in cycles to failure; $n = 15$ pumps were tested for a total time of $t = 15,962,989$. Eight failures ($s = 8$) were observed and the smallest failure time was $t_1 = 237,217$.

a) Suppose one transforms the parameters to the real line by the transformations $\theta_1 = \log \beta, \theta_2 = \log(t_1 - \mu)$. Write down the posterior density of (θ_1, θ_2).

b) Construct an R function that computes the log posterior density of (θ_1, θ_2).

c) Use the `laplace` function to approximate the posterior density.

d) Use a multivariate t proposal density and the SIR algorithm to simulate a sample of 1000 draws from the posterior distribution.

e) Suppose one is interested in estimating the reliability at time t_0 defined by

$$R(t_0) = e^{-(t_0 - \mu)/\beta}.$$

Using your simulated values from the posterior, find the posterior mean and posterior standard deviation of $R(t_0)$ when $t_0 = 10^6$ cycles.

4. **Poisson regression**

Haberman (1978) considers an experiment involving subjects reporting one stressful event. The collected data are $y_1, ..., y_{18}$, where y_i is the number of events recalled i months before the interview. Suppose y_i is distributed Poisson with mean λ_i, where the $\{\lambda_i\}$ satisfy the loglinear regression model

$$\log \lambda_i = \beta_0 + \beta_1 i.$$

The data are shown in Table 5.2. If (β_0, β_1) is assigned a uniform prior, then the logarithm of the posterior density is given, up to an additive constant, by

$$\log g(\beta_0, \beta_1 | \text{data}) = \sum_{i=1}^{18} \left[y_i(\beta_0 + \beta_1 i) - \exp(\beta_0 + \beta_1 i) \right].$$

Table 5.2. Numbers of subjects recalling one stressful event.

Months	1	2	3	4	5	6	7	8	9	10	11	12	13	14	15	16	17	18
y_i	15	11	14	17	5	11	10	4	8	10	7	9	11	3	6	1	1	4

a) Write a R function to compute the logarithm of the posterior density of (β_0, β_1).

b) Suppose we are interested in estimating the posterior mean and standard deviation for the slope β_1. Approximate these moments by a normal approximation about the posterior mode (function `laplace`).

c) Use a multivariate t proposal density and the SIR algorithm to simulate 1000 draws from the posterior density. Use this sample to estimate the posterior mean and standard deviation of the slope β_1. Compare your estimates with the estimates using the normal approximation.

5. **Grouped Poisson data**

Hartley (1958) fits a Poisson model to the following grouped data:

Number of Events	0	1	2	3+	Total
Group Frequency	11	37	64	128	240

Suppose the mean Poisson parameter is λ and the frequency of observations with j events is $n_j, j = 0, 1, 2$, and n_3 is the frequency of observations with at least three events. If the standard noninformative prior $g(\lambda) = 1/\lambda$ is assigned, then the posterior density is given by

$$g(\lambda|\text{data}) \propto e^{-\lambda(n_0+n_1+n_2)} \lambda^{n_1+2n_2-1} \left[1 - e^{-\lambda}\left(1 + \lambda + \frac{\lambda^2}{2}\right)\right]^{n_3}.$$

• Write an R function to compute the logarithm of the posterior density of λ.

• Use the function `laplace` to find a normal approximation to the posterior density of the transformed parameter $\theta = \log \lambda$.

• Use a t proposal density and the SIR algorithm to simulate 1000 draws from the posterior. Use the simulated sample to estimate the posterior mean and standard deviation of λ. Compare the estimates with the normal approximation estimates found in part (a).

6

Markov Chain Monte Carlo Methods

6.1 Introduction

In Chapter 5, we introduced the use of simulation in Bayesian inference. Rejection sampling is a general method for simulating from an arbitrary posterior distribution, but it can be difficult to set up since it requires the construction of a suitable proposal density. Importance sampling and SIR algorithms are also general-purpose algorithms, but they also require proposal densities that may be difficult to find for high-dimensional problems. In this chapter, we illustrate the use of Markov chain Monte Carlo (MCMC) algorithms in summarizing posterior distributions. Markov chains are introduced in the discrete state space situation in Section 6.2. Through a simple random walk example, we illustrate some of the important properties of a special Markov chain, and we use R to simulate from the chain and move toward the stationary distribution. In Section 6.3, we describe two variants of the popular Metropolis-Hastings algorithms in setting up Markov chains, and in Section 6.4, we describe Gibbs sampling where the Markov chain is set up through the conditional distributions of the posterior. We describe one strategy for summarizing a posterior distribution and illustrate it for three problems. MCMC algorithms are very attractive in that they are easy to set up and program and require relatively little prior input from the user. R is a convenient language for programming these algorithm and is also very suitable for performing output analysis, where one does several graphical and numerical computations to check if the algorithm is indeed producing draws from the target posterior distribution.

6.2 Introduction to Discrete Markov Chains

Suppose a person takes a random walk on a number line on the values 1, 2, 3, 4, 5, 6. If the person is currently at an interior value (2, 3, 4, or 5), in the next second she is equally likely to remain at that number or move to an adjacent number. If she does move, she is equally likely to move left or right. If the person is currently at one of the end values (1 or 6), in the next second she is equally likely to stay or move to the adjacent location.

This is a simple example of a discrete Markov chain. A Markov chain describes probabilistic movement between a number of states. Here there are six possible states, 1 through 6, corresponding to the possible location of the walker. Given that the person is at a current location, she moves to other locations with specified probabilities. The probability she moves to another location depends only on her current location and not on previous locations visited. We describe movement between states in terms of transition probabilities – they describe the likelihoods of moving between all possible states in a single step in a Markov chain. We summarize the transition probabilities by means of a transition matrix T:

$$
T = \begin{bmatrix}
.50 & .50 & 0 & 0 & 0 & 0 \\
.25 & .50 & .25 & 0 & 0 & 0 \\
0 & .25 & .50 & .25 & 0 & 0 \\
0 & 0 & .25 & .50 & .25 & 0 \\
0 & 0 & 0 & .25 & .50 & .25 \\
0 & 0 & 0 & 0 & .50 & .50
\end{bmatrix}
$$

The first row in T gives the probabilities of moving to all states 1 through 6 in a single step from location 1, the second row gives the transition probabilities in a single step from location 2, and so on.

There are several important properties of this particular Markov chain. It is possible to go from every state to every state in one or more steps – a Markov chain with this property is said to be *irreducible*. Given that the person is in a particular state, if the person can only return to this state at regular intervals, then the Markov chain is said to be *periodic*. This example is *aperiodic* since it is not a periodic Markov chain.

We can represent one's current location as a probability row vector of the form

$$
p = (p_1, p_2, p_3, p_4, p_5, p_6),
$$

where p_i represents the probability the person is currently in state i. If p^j represents the location of the traveler at step j, then the location of the traveler at the $j + 1$ step is given by the matrix product

$$
p^{j+1} = p^j T.
$$

Suppose we can find a probability vector w such that $wP = w$. Then w is said to be the *stationary* distribution. If a Markov chain is irreducible and aperiodic, then it has a unique stationary distribution. Moreover, the limiting distribution of this Markov chain, as the number of steps approaches infinity, will be equal to this stationary distribution.

We can empirically demonstrate the existence of the stationary distribution of our Markov chain by running a simulation experiment. We start our random walk at a particular state, say location 3, and then simulate many steps of the Markov chain using the transition matrix T. The relative frequencies of our traveler in the six locations after many steps will eventually approach the stationary distribution w.

We start our simulation in R by reading in the transition matrix T and setting up a storage vector s for the locations of our traveler in the random walk.

```
> T=matrix(c(.5,.5,0,0,0,0,.25,.5,.25,0,0,0,0,.25,.5,.25,0,0,
+            0,0,.25,.5,.25,0,0,0,0,.25,.5,.25,0,0,0,0,.5,.5),
+ nrow=6,ncol=6,byrow=TRUE)
> T

     [,1] [,2] [,3] [,4] [,5] [,6]
[1,] 0.50 0.50 0.00 0.00 0.00 0.00
[2,] 0.25 0.50 0.25 0.00 0.00 0.00
[3,] 0.00 0.25 0.50 0.25 0.00 0.00
[4,] 0.00 0.00 0.25 0.50 0.25 0.00
[5,] 0.00 0.00 0.00 0.25 0.50 0.25
[6,] 0.00 0.00 0.00 0.00 0.50 0.50

> s=array(0,c(50000,1))
```

We indicate that the starting location for our traveler is state 3 and perform a loop to simulate 50,000 draws from the Markov chain. We use the `sample` function to simulate one step – the arguments to this function indicate that we are sampling a single value from the set $\{1, 2, 3, 4, 5, 6\}$ with probabilities given by the s^{j-1} row of the transition matrix T, where s^{j-1} is the current location of our traveler.

```
> s[1]=3
> for (j in 2:50000)
+   s[j]=sample(1:6,size=1,prob=T[s[j-1],])
```

We summarize the frequencies of visits to the six states after 500, 2000, 8000, and 50,000 steps of the chain by use of the `table` command; we convert the counts to relative frequencies by dividing by the number of steps.

```
> m=c(500,2000,8000,50000)
> for (i in 1:4)
+   print(table(s[1:m[i]])/m[i])


    1     2     3     4     5     6
0.164 0.252 0.174 0.130 0.174 0.106


     1      2      3      4      5      6
0.1205 0.1965 0.1730 0.1735 0.2170 0.1195
```

```
     1        2        3        4        5        6
0.109250 0.188000 0.183875 0.194625 0.212000 0.112250
```

```
    1       2       3       4       5       6
0.10970 0.20770 0.19450 0.19342 0.19628 0.09840
```

It appears from the output that the relative frequencies of the states are converging to the stationary distribution $w = (0.1, 0.2, 0.2, 0.2, 0.2, 0.1)$. We can confirm that w is indeed the stationary distribution of this chain by multiplying w by the transition matrix T:

```
> w=matrix(c(.1,.2,.2,.2,.2,.1),nrow=1,ncol=6)
> w%*%T
```

```
     [,1] [,2] [,3] [,4] [,5] [,6]
[1,]  0.1  0.2  0.2  0.2  0.2  0.1
```

6.3 Metropolis-Hasting Algorithms

A popular way of simulating from a general posterior distribution is by Markov chain Monte Carlo (MCMC) methods. This essentially is a continuous-valued generalization of the discrete Markov chain setup described in the previous section. The MCMC sampling strategy sets up an irreducible, aperiodic Markov chain for which the stationary distribution equals the posterior distribution of interest. A general way of constructing a Markov chain is by a Metropolis-Hastings algorithm. In this section, we focus on two particular variants of Metropolis-Hastings algorithms, the independence chain and the random walk chain, that are applicable to a wide variety of Bayesian inference problems.

Suppose we wish to simulate from a posterior density $g(\theta|y)$. In the following, to simplify notation, we write the density simply as $g(\theta)$. A Metropolis-Hastings algorithm begins with an initial value θ^0 and specifies a rule for simulating the tth value in the sequence θ^t given the $(t-1)$st value in the sequence θ^{t-1}. This rule consists of a *proposal density* which simulates a candidate value θ^*, and the computation of an *acceptance probability* P that indicates the probability the candidate value will be accepted to be the next value in the sequence. Specifically, this algorithm can be described as follows:

- Simulate a candidate value θ^* from a proposal density $p(\theta^*|\theta^{t-1})$.
- Compute the ratio
$$R = \frac{g(\theta^*)p(\theta^{t-1}|\theta^*)}{g(\theta^{t-1})p(\theta^*|\theta^{t-1})}.$$
- Compute the acceptance probability $P = \min\{R, 1\}$.
- Sample a value θ^t such that $\theta^t = \theta^*$ with probability P; otherwise $\theta^t = \theta^{t-1}$.

Under some easily satisfied regularity conditions on the proposal density $p(\theta^*|\theta^{t-1})$, the sequence of simulated draws $\theta^1, \theta^2, \dots$ will converge to a random variable that is distributed according to the posterior distribution $g(\theta)$.

Different Metropolis-Hastings algorithms are constructed by the choice of proposal density. If the proposal density is independent of the current value in the sequence, that is,

$$p(\theta^*|\theta^{t-1}) = p(\theta^*),$$

then the resulting algorithm is called an *independence* chain. Other proposal densities can be defined by letting the density have the form

$$p(\theta^*|\theta^{t-1}) = h(\theta^* - \theta^{t-1}),$$

where h is a symmetric density about the origin. In this type of *random walk* chain, the ratio R has the simple form

$$R = \frac{g(\theta^*)}{g(\theta^{t-1})}.$$

The R functions `rwmetrop` and `indepmetrop` in the LearnBayes package implement, respectively, the random-walk and independence Metropolis-Hasting algorithms for special choices of proposal densities. For the function `rwmetrop`, the proposal density has the form

$$\theta^* = \theta^{t-1} + \text{scale } Z,$$

where Z is multivariate normal with mean vector 0 and variance-covariance matrix V and *scale* is a positive scale parameter. For the function `indepmetrop`, the proposal density for θ^* is multivariate normal with mean vector μ and covariance matrix V.

To use a Metropolis-Hastings algorithm, one first decides on the proposal density and then obtains a simulated sample of draws $\{\theta^t, t = 1, \dots m\}$ by use of the R functions `rwmetrop` or `indepmetrop`. The output of each of these functions has two components: `par` is a matrix of simulated draws where each row corresponds to a value of θ, and `accept` gives the acceptance rate of the algorithm.

Desirable features of the proposal density in an algorithm depend on the MCMC algorithm employed. For an independence chain, we desire that the proposal density p approximates the posterior density g, suggesting a high acceptance rate. But, as in rejection sampling, it is important that the ratio g/p is bounded, especially in the tail portion of the posterior density. This means that one may choose a proposal p that is more diffuse than the posterior, resulting in a lower acceptance rate. For random walk chains with normal proposal densities, it has been suggested that acceptance rates between 25% and 45% are good. The "best" choice of acceptance rate ranges from 45% for one and two parameters to 25% for problems with more parameters. This advice also applies when one monitors the Metropolis within Gibbs algorithm described in Section 6.4.

6.4 Gibbs Sampling

One of the attractive methods of setting up an MCMC algorithm is Gibbs sampling. Suppose that the parameter vector of interest is $\theta = (\theta_1, ..., \theta_p)$. The joint posterior distribution of θ, which we denote by $[\theta|\text{data}]$, may be of high dimension and difficult to summarize. Suppose we define the set of conditional distributions

$$[\theta_1|\theta_2, ..., \theta_p, \text{data}],$$

$$[\theta_2|\theta_1, \theta_3, ..., \theta_p, \text{data}],$$

$$...$$

$$[\theta_p|\theta_1, ..., \theta_{p-1}, \text{data}],$$

where $[X|Y, Z]$ represents the distribution of X conditional on values of the random variables Y and Z. The idea behind Gibbs sampling is that we can set up a Markov chain simulation algorithm from the joint posterior distribution by successfully simulating individual parameters from the set of p conditional distributions. Simulating in turn one value of each individual parameter from these distributions is called one cycle of Gibbs sampling. Under general conditions, draws from this simulation algorithm will converge to the target distribution (the joint posterior of θ) of interest.

In situations where it is not convenient to sample directly from the conditional distributions, one can use a Metropolis algorithm such as the random walk type to simulate from each distribution. A "Metropolis within Gibbs" algorithm of this type is programmed in the function `gibbs` in the LearnBayes package. Suppose that θ_i^t represents the current value of θ_i in the simulation and let $g(\theta_i)$ represent the conditional distribution where we have suppressed the dependence of this distribution on values of the remaining components of θ. Then a candidate value for θ_i is given by

$$\theta_i^* = \theta_i^t + c_i Z,$$

where Z is a standard normal variate and c_i is a fixed scale parameter. The next simulated value of θ_i, θ_i^{t+1}, will be equal to the candidate value with probability $P = \min\{1, g(\theta_i^*)/g(\theta_i^t)\}$; otherwise the value $\theta_i^{t+1} = \theta_i^t$. To use the function `gibbs`, one inputs the function defining the log posterior, the starting value of the simulation, the number of Gibbs cycles, and a vector of scale parameters containing $c_1, ..., c_p$. The output of `gibbs` is a list; the component `par` is a matrix of simulated draws and `accept` is a vector of acceptance rates for the individual Metropolis steps.

6.5 MCMC Output Analysis

For the MCMC algorithms described in this book, the distribution of the simulated value at the jth iterate, θ^j, will converge to a draw from the posterior distribution as j approaches infinity. Unfortunately, this theoretical result provides no practical guidance on how to decide if the simulated sample provides a reasonable approximation to the posterior density $g(\theta|\text{data})$.

In typical practice, one monitors the performance of an MCMC algorithm by inspecting the value of the acceptance rate, constructing graphs, and computing diagnostic statistics on the stream of simulated draws. We call this investigation an *MCMC output analysis*. By means of this exploratory analysis, one decides if the chain has sufficiently explored the entire posterior distribution (there is good *mixing*) and the sequence of draws has approximately converged. If one has a sample from the posterior distribution, then one wishes to obtain a sufficient number of draws so that one can accurately estimate any particular summary of the posterior of interest.

In this section we briefly describe some of the important issues in interpreting MCMC output and describe a few graphical and numerical diagnostics for assessing convergence. One issue in understanding MCMC output is detecting the size of the burn-in period. The simulated values of θ obtained at the beginning of an MCMC run are not distributed from the posterior distribution. However, after some number of iterations have been performed (the burn-in period), the effect of the initial values wears off and the distribution of the new iterates approaches the true posterior distribution. One way of estimating the length of the burn-in period is to examine *trace plots* of simulated values of a component or particular function of θ against the iteration number. Trace plots are especially important when MCMC algorithms are initialized with parameter values that are far from the center of the posterior distribution.

A second concern in analyzing output from MCMC algorithms is the degree of autocorrelation in the sampled values. In both the Metropolis and Gibbs sampling algorithms, the simulated value of θ at the $(j + 1)$st iteration is dependent on the simulated value at the jth iteration. If there is strong correlation between successive values in the chain, then two consecutive values provide only marginally more information about the posterior distribution than a single simulated draw. Also, a strong correlation between successive iterates may prevent the algorithm from exploring the entire region of the parameter space. A standard statistic for measuring the degree of dependence between successive draws in the chain is the autocorrelation that measures the correlation between the sets $\{\theta^j\}$ and $\{\theta^{j+L}\}$, where L is the lag or number of iterates separating the two sets of values. A standard graph is to plot the values of the autocorrelation against the log L. If the chain is mixing adequately, the values of the autocorrelation will decrease to zero as the lag value is increased.

Another issue that arises in output analysis is the choice of the simulated sample size and the resulting accuracy of calculated posterior summaries. Since iterates in an MCMC algorithm are not independent, one cannot use standard "independent sample" methods to compute estimated standard errors. One simple method of computing standard errors for correlated output is the method of batch means. Suppose we estimate the posterior mean of θ_i with the summary sample mean

$$\bar{\theta}_i = \frac{\sum_{j=1}^{m} \theta_i^j}{m}.$$

What is the simulation standard error of this estimate? In the batch means method, the stream of simulated draws $\{\theta_i^j\}$ is subdivided into b batches, each batch of size v, where $m = bv$. In each batch, we compute a sample mean; call the set of sample means $\bar{\theta}_i^1, ..., \bar{\theta}_i^b$. If the lag one autocorrelation in the sequence in the batch means is small, then we can approximate the standard error of the estimate $\bar{\theta}_i$ by the standard deviation of the batch means divided by the square root of the number of batches.

6.6 A Strategy in Bayesian Computing

For a particular Bayesian inference problem, we assume that one has defined the log posterior density by an R function. Following the recommendation of Gelman et al (2003), Chapter 11, a good approach for summarizing this density is to set up a Markov chain simulation algorithm. The Metropolis-Hastings random walk and independence chains and the Gibbs sampling algorithm are attractive Markov chains since they are easy to program and require relatively little prior input. But these algorithms do require some initial guesses at the location and spread of the parameter vector θ. These initial guesses can be found by non-Bayesian methods such as the method of moments or maximum likelihood. Alternatively, one can obtain an approximation to the posterior distribution by finding the mode by use of some optimization algorithm. For example, Newton's method gives the posterior mode and an approximation to the variance-covariance matrix that can be used in specifying the proposal densities in the Metropolis-Hastings algorithms.

In our examples, we illustrate the use of the function `laplace` to locate the posterior density. We can check the accuracy of the normal approximation in the two-parameter case by the construction of a contour graph of the joint posterior. These examples show that there can be some errors in the normal approximation. But the `laplace` function is still helpful in that the values of $\hat{\theta}$ and V can be used to construct efficient Metropolis-Hastings algorithms for simulating from the exact joint posterior distribution. Once one has decided that the simulated stream of values represents an approximate sample from the posterior, then one can summarize this sample in different ways to perform inferences about θ.

6.7 Learning About a Normal Population from Grouped Data

As a first example, suppose a random sample is taken from a normal population with mean μ and standard deviation σ. But one only observes the data in "grouped" form, where the frequencies of the data in bins are recorded. For example, suppose one is interested in learning about the mean and standard deviation of the heights (in inches) of men from a local college. One is given the summary frequency data shown in Table 6.1. One sees that 14 men were shorter than 66 inches, 30 men had heights between 66 and 68 inches, and so on.

Table 6.1. Grouped frequency data for heights of male students at a college.

Height Interval (in.)	Frequency
less than 66	14
between 66 and 68	30
between 68 and 70	49
between 70 and 72	70
between 72 and 74	33
over 74	15

We are observing multinomial data with unknown bin probabilities $p_1, ..., p_6$ where the probabilities are functions of the unknown parameters of the normal population. For example, the probability that a student's height is between 66 and 68 inches is given by $p_2 = \Phi(68, \mu, \sigma) - \Phi(66, \mu, \sigma)$, where $\Phi(; \mu, \sigma)$ is the cdf of a normal(μ, σ) random variable. It is straightforward to show that the likelihood of the normal parameters given this grouped data is given by

$$L(\mu, \sigma) = \Phi(66, \mu, \sigma)^{14}(\Phi(68, \mu, \sigma) - \Phi(66, \mu, \sigma))^{30}$$
$$\times (\Phi(70, \mu, \sigma) - \Phi(68, \mu, \sigma))^{49}(\Phi(72, \mu, \sigma) - \Phi(70, \mu, \sigma))^{70}$$
$$\times (\Phi(74, \mu, \sigma) - \Phi(72, \mu, \sigma))^{33}(1 - \Phi(74, \mu, \sigma))^{15}.$$

Suppose (μ, σ) are assigned the usual noninformative prior proportional to $1/\sigma$. Then the posterior density of the parameters is proportional to

$$g(\mu, \sigma | \text{data}) \propto \frac{1}{\sigma}L(\mu, \sigma).$$

Following our general strategy, we transform the positive standard deviation by $\lambda = \log(\sigma)$ and the posterior density of (μ, λ) is given by

$$g(\mu, \lambda | \text{data}) \propto L(\mu, \exp(\lambda)).$$

We begin by writing a short function groupeddatapost that computes the logarithm of the posterior density of (μ, λ). There are two arguments to this function: a matrix theta, where each row corresponds to a value of (μ, λ), and a list data. The list has two components: data$b is a vector of cutpoints for the bins and data$f is a vector of bin frequencies.

```
groupeddatapost=function(theta,data)
{
cpoints=data$b
freq=data$f
nbins=length(cpoints)
m=theta[,1]; logsigma=theta[,2]
z=0*m; s=exp(logsigma)
z=freq[1]*log(pnorm(cpoints[1],m,s))
for (j in 1:(nbins-1))
  z=z+freq[j+1]*log(pnorm(cpoints[j+1],m,s)-
    pnorm(cpoints[j],m,s))
z=z+freq[nbins]*log(1-pnorm(cpoints[nbins],m,s))
return(z)
}
```

We begin by defining the grouped data by the list d.

```
> d=list(b=seq(66,74,by=2),f=c(14,30,49,70,33,15))
```

To use the function laplace, one requires a good guess at the location of the posterior mode. To estimate the mode of $(\mu, \log \sigma)$, we first create an artificial continuous dataset by replacing each grouped observation by its bin midpoint. Then we approximate the posterior mode by computing the sample mean and the logarithm of the standard deviation of these artificial observations.

```
> y=c(rep(65,14),rep(67,30),rep(69,49),rep(71,70),rep(73,33),
+   rep(75,15))
> mean(y)
```

```
[1] 70.16588
```

```
> log(sd(y))
```

```
[1] 0.9504117
```

Based on this computation, we believe that the posterior of the vector $(\mu, \log \sigma)$ is approximately (70, 1). We use the laplace function, where the log posterior is defined in the function groupeddatapost, start is set equal to this starting value, 10 iterations of Newton's method are run, and the grouped data are contained in the list d.

```
> start=array(c(70,1),c(1,2))
> fit=laplace(groupeddatapost,start,10,d)
> fit
```

```
$mode
           [,1]      [,2]
[1,]  70.61358  1.104264
```

```
$var
             [,1]            [,2]
[1,]  0.0425681264  0.0005563627
[2,]  0.0005563627  0.0032063738
```

```
$int
[1] -391.794
```

From the output, the posterior mode of $(\mu, \log \sigma)$ is found to be $(70.61, 1.10)$. The associated posterior standard deviations of the parameters can be estimated by computing the square roots of the diagonal elements of the variance-covariance matrix.

```
> modal.sds=sqrt(diag(fit$var))
```

We use the output from the function `laplace` to design a Metropolis random walk algorithm to simulate from the joint posterior. For the proposal density we use the variance-covariance matrix obtained from `laplace` and we set the scale parameter equal to 2. We run 10,000 iterations of the random walk algorithm starting at the value `start`. The output `fit2` is a list with two components: `par` is a matrix of simulated values where each row corresponds to a single draw of the parameter vector, and `accept` gives the acceptance rate of the random walk chain.

```
> proposal=list(var=fit$var,scale=2)
> fit2=rwmetrop(groupeddatapost,proposal,start,10000,d)
```

We monitor the algorithm by displaying the acceptance rate; here the value is .2937 which is close to the desired acceptance rate for this Metropolis random walk algorithm.

```
> fit2$accept
```

```
[1] 0.2937
```

We can summarize the parameters μ and $\log \sigma$ by computation of the posterior means and posterior standard deviations.

```
> post.means=apply(fit2$par,2,mean)
> post.sds=apply(fit2$par,2,sd)
```

One can assess the accuracy of the model approximation to the posterior by comparing the means and standard deviations from the function `laplace` with the values computed from the simulated output from the MCMC algorithm.

```
> cbind(c(fit$mode),modal.sds)
```

```
                 modal.sds
[1,] 70.613579  0.20632045
[2,]  1.104264  0.05662485
```

```
> cbind(post.means,post.sds)

      post.means    post.sds
[1,]   70.609472 0.21042984
[2,]    1.110115 0.05790169
```

For this model, there is close agreement in the two sets of posterior moments which indicates that the modal approximation to the posterior distribution is reasonably accurate.

We confirm this statement by using the function mycontour to draw a contour plot of the joint posterior of μ and $\log \sigma$. The last 5000 simulated draws from the random walk Metropolis algorithm are drawn on top in Fig. 6.1. Note that the contour lines have an elliptical shape that confirms the accuracy of the normal approximation in this example.

```
> mycontour(groupeddatapost,c(69.5,71.5,.8,1.4),d)
> points(fit2$par[5001:10000,1],fit2$par[5001:10000,2])
> title(xlab="mu",ylab="log sigma")
```

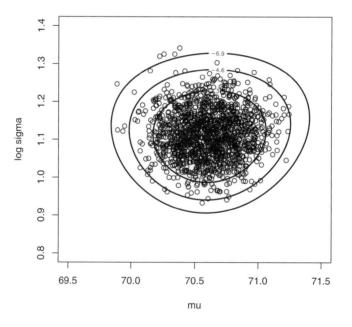

Fig. 6.1. Contour plot of posterior of μ and $\log \sigma$ for grouped data example. A simulated sample of 5000 draws of the posterior is also shown.

6.8 Example of Output Analysis

We illustrate the use of MCMC output analysis by use of the R package boa that will be described in Chapter 11. Suppose we rerun the Metropolis random walk algorithm for the grouped data posterior with poor choices of starting value and proposal density. As a starting value, we choose $(\mu, \log \sigma) = (65, 1)$ (the choice of μ is too small) and we select the small scale factor of 0.2 (instead of 2):

```
> start=array(c(65,1),c(1,2))
> proposal=list(var=fit$var,scale=0.2)
```

We then rerun the Metropolis function rwmetrop:

```
> bayesfit=rwmetrop(groupeddatapost,proposal,start,10000,d)
```

We find that the acceptance rate of this modified algorithm is 0.89 which is much larger than the 0.29 rate that we found using the scale factor 2.

Fig. 6.2 displays a trace plot of the simulated draws of μ from this Metropolis algorithm. Note that there is a significant burn-in period, approximately 600 iterations, before the simulated draws reach the main support of the posterior of μ. Also note the irregularity of the simulated sequence; for example, the iterates will explore the region where $\mu > 71$ for a while before returning to the center of the distribution.

One can observe the strong correlation structure of the sequence by the use of an autocorrelation plot shown in Fig. 6.3. The autocorrelations are close to one for lag one and reduce very slowly as a function of the lag.

The following summary output of the simulated draws of μ confirm the behavior of the MCMC run seen in Fig. 6.2 and Fig. 6.3. The estimate at the posterior mean of μ is 70.32. If we assume naively that this simulated sample represented independent draws, then the standard error of this estimate is .0106. However, a more accurate estimate at the standard error is the Batch SE given by .0744. The lag one autocorrelation of the batch means (using batches of size 50) is .924.

```
SUMMARY STATISTICS:
====================
Bin size for calculating Batch SE and (Lag 1) ACF = 50

Chain: mu
```

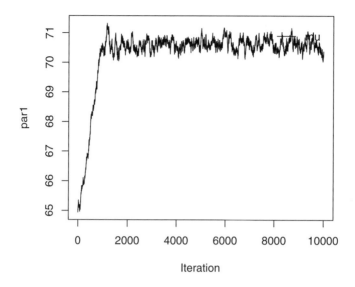

Fig. 6.2. Trace plot of simulated draws of μ for an MCMC chain with poor choices for starting value and scale factor.

	Mean	SD	Naive SE	MC Error	Batch SE	Batch ACF
par1	70.31933	1.055536	0.01055536	0.1395873	0.07443487	0.9243932
	0.025	0.5	0.975	MinIter	MaxIter	Sample
par1	65.98562	70.57855	70.99605	1	10000	10000

It is instructive to compare these diagnostic graphs with the graphs using the better starting value and choice of proposal density used in Section 6.7. Fig. 6.4 and Fig. 6.5 display a trace plot and autocorrelation graph of the simulated draws of μ using the starting value $(\mu, \log \sigma) = (70, 1)$ and scale factor equal to 2. The trace plot of the simulated stream of μ looks more like random noise. The lag one autocorrelation is high, but the autocorrelation values dissipate rapidly as a function of the lag.

As before, we can compute summary statistics for this stream of MCMC output.

Sampler Lag–Autocorrelations

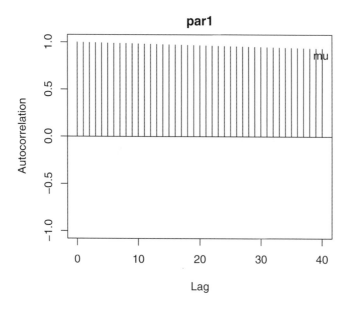

Fig. 6.3. Autocorrelation plot of simulated draws of μ for an MCMC chain with poor choices for starting value and scale factor.

```
SUMMARY STATISTICS:
====================
Bin size for calculating Batch SE and (Lag 1) ACF = 50

Chain: mu
---------
          Mean           SD      Naive SE     MC Error      Batch SE
par1  70.61206   0.2065044  0.002065044  0.005924462   0.005778565
       Batch ACF       0.025         0.5        0.975  MinIter  MaxIter
par1  -0.003788097   70.19776   70.61043   71.01351        1    10000
        Sample
par1     10000
```

Here the estimate of the posterior mean of μ is 70.61 with a batch standard error of .006. The autocorrelation between batch means of size 50 is the small value -0.0037. The graphs and the summary statistics confirm the better performance of the MCMC chain with a starting value $(\mu, \log \sigma) = (70, 1)$ and scale factor of 2.

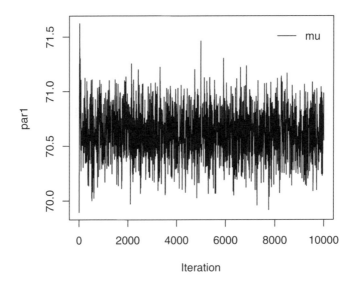

Fig. 6.4. Trace plot of simulated draws of μ for MCMC chain with good choices for starting value and scale factor.

6.9 Modeling Data with Cauchy Errors

For a second example, suppose that we are interested in modeling data where outliers may be presented. Suppose $y_1, ..., y_n$ are a random sample from a Cauchy density with location parameter μ and scale parameter σ:

$$f(y|\mu, \sigma) = \frac{1}{\pi\sigma(1 + z^2)},$$

where $z = (y - \mu)/\sigma$. Suppose that we assign the usual noninformative prior to (μ, σ):

$$g(\mu, \sigma) = \frac{1}{\sigma}.$$

The posterior density of μ and σ is given, up to a proportionality constant, by

$$g(\mu, \sigma|\text{data}) \propto \frac{1}{\sigma} \prod_{i=1}^{n} f(y_i|\mu, \sigma).$$

$$= \frac{1}{\sigma} \prod_{i=1}^{n} \left[\frac{1}{\sigma} \left(1 + (y_i - \mu)^2/\sigma^2 \right)^{-1} \right].$$

Sampler Lag–Autocorrelations

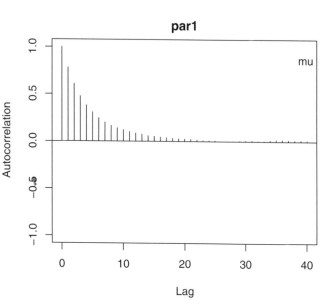

Fig. 6.5. Autocorrelation plot of simulated draws of μ for an MCMC chain with good choices for starting value and scale factor.

Again we first transform the positive parameter σ to the real line by the reexpression $\lambda = \log \sigma$, leading to the posterior density of (μ, λ):

$$g(\mu, \lambda | \text{data}) \propto \prod_{i=1}^{n} \left[\exp(-\lambda) \left(1 + \exp(-2\lambda)(y_i - \mu)^2 \right)^{-1} \right].$$

The logarithm of the density is then given, up to an additive constant, by

$$\log g(\mu, \lambda | \text{data}) = \sum_{i=1}^{n} \left[-\lambda - \log \left(1 + \exp(-2\lambda)(y_i - \mu)^2 \right) \right].$$

We write the following R function `cauchyerrorpost` to compute the logarithm of the posterior density. There are two arguments to the function: `theta`, a matrix where each row corresponds to a value of the pair (μ, λ), and the vector of observations `y`. Since the parameters are vectors, we use a loop in the function where the individual terms of the log likelihood are summed over the values of $y_1, ..., y_n$. To simplify the code, we use the R function `dt`, which computes the density of the t random variable. (The Cauchy density is the t density with a single degree of freedom.)

```
cauchyerrorpost=function(theta,y)
{
mu=theta[,1]; lambda=theta[,2]
sigma=exp(lambda)
val=0*mu
for (i in 1:length(y))
    {val=val+log(dt((y[i]-mu)/sigma,df=1)/sigma)}
return(val)
}
```

We apply this model to Darwin's famous dataset concerning 15 differences of the heights of cross- and self-fertilized plants quoted by Fisher (1960). This dataset can be found in the LearnBayes library with the name darwin. We read in the dataset and attach the dataframe so we can access the variable difference. We initially compute the mean and logarithm of the standard deviation of the data to get some initial estimates at the locations of the posterior distributions of μ and $\lambda = \log(\sigma)$.

```
> data(darwin)
> attach(darwin)
> mean(difference)

[1] 21.66667

> log(sd(difference))

[1] 3.65253
```

To find the posterior mode, we use the function laplace. The arguments are the name of the function cauchyerrorpost defining the log posterior density, an array of initial estimates at the parameters, the number of iterations of the Newton-Raphson algorithm, and the data used in the log posterior function. For initial estimates, we use the values $\mu = 21.6, \lambda = 3.6$ found earlier, and we use 10 iterations of the algorithm.

```
> laplace(cauchyerrorpost,array(c(21.6,3.6),c(1,2)),10,difference)

$mode
          [,1]      [,2]
[1,] 24.70160 2.772829

$var
              [,1]         [,2]
[1,] 34.9647321 0.3672069
[2,]  0.3672069 0.1378207

$int
[1] -73.24035
```

The posterior mode is given by $(\mu, \lambda) = (24.7, 2.77)$. The output also gives the associated variance-covariance matrix and an estimate at the log integral.

We can use these estimates of center and spread to construct a rectangle that covers essentially all of the posterior probability of the parameters. As an initial guess at this rectangle, we take for each parameter the posterior mode plus and minus four standard deviations, where the standard deviations are obtainable from the diagonal elements of the variance-covariance matrix.

```
> c(24.7-4*sqrt(34.96),24.7+4*sqrt(34.96))
```

[1]   1.049207 48.350793

```
> c(2.77-4*sqrt(.138),2.77+4*sqrt(.138))
```

[1] 1.284066 4.255934

After some trial and error, we use the rectangle $\mu \in (-10, 60)$, $\lambda \in (1, 4.5)$ as the bounding rectangle for the function `mycontour`. Fig. 6.6 displays the contour graph of the exact posterior distribution.

```
> mycontour(cauchyerrorpost,c(-10,60,1,4.5),difference)
> title(xlab="mu",ylab="log sigma")
```

The contours of the exact posterior distribution have an interesting shape and one may wonder how these contours compare to those for a bivariate normal approximation. In the R code, we rerun the `laplace` function to obtain the posterior mode `t$mode` and associated variance-covariance matrix `t$var`. Using these values as inputs, we draw contours of a bivariate normal density in Fig. 6.7 where the log bivariate normal density is programmed in the function `lbinorm`. The eliptical shape of these normal contours seems significantly different from the shape of the exact posterior contours, which indicates that the normal approximation may be inadequate.

```
> fitlaplace=laplace(cauchyerrorpost,array(c(21.6,3.6),c(1,2)),
+    10,difference)
> mycontour(lbinorm,c(-10,60,1,4.5),list(m=fitlaplace$mode,
+    v=fitlaplace$var))
> title(xlab="mu",ylab="log sigma")
```

Although the normal approximation may not be the best summary of the posterior distribution, the estimated variance-covariance matrix is helpful in setting up a Metropolis random walk chain. We initially define a list `proposal` that contains the estimated variance-covariance matrix and a scale factor. We define the starting value of the chain in the array `start`. The simulation algorithm is run using the function `rwmetrop`. The inputs are the function defining the log posterior, the list `proposal`, the starting value, the number of simulations, and the data vector.

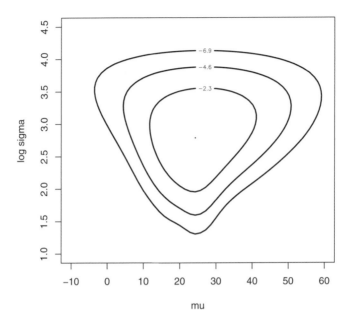

Fig. 6.6. Contour plot of posterior of μ and $\log \sigma$ for Cauchy error model problem.

```
> proposal=list(var=fitlaplace$var,scale=2.5)
> start=array(c(20,3),c(1,2))
> m=1000
> s=rwmetrop(cauchyerrorpost,proposal,start,m,difference)
> mycontour(cauchyerrorpost,c(-10,60,1,4.5),difference)
> title(xlab="mu",ylab="log sigma")
> points(s$par[,1],s$par[,2])
```

In Fig. 6.8 we display simulated draws from `rwmetrop` on top of the contour graph.

Fig. 6.9 and Fig. 6.10 show the "exact" marginal posterior densities of μ and $\log \sigma$ found from a density estimate from 50,000 simulated draws from the random walk algorithm. Also each figure shows the approximate normal approximation from the `laplace` output. These figures demonstrate the non-normal shape of these marginal posteriors.

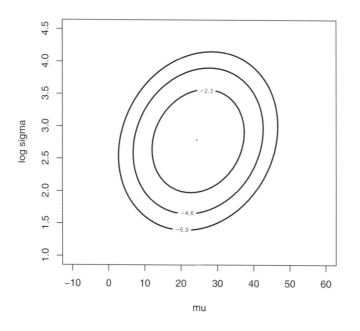

Fig. 6.7. Contour plot of normal approximation to posterior of μ and $\log \sigma$ for Cauchy error model problem.

It is instructive to illustrate "brute-force" and other Metropolis-Hastings algorithms for this problem. The brute-force algorithm is based on simulating draws of $(\mu, \log \sigma)$ from the grid using the function `simcontour`. One can use a Metropolis-Hastings independence chain, where the proposal density is multivariate normal with mean and variance given by the normal approximation. Alternatively, one can apply a Gibbs sampling algorithm with a vector of scale parameters equal to $(12, .75)$; these values are approximately equal to twice the estimated posterior standard deviations of the two parameters. All the simulation algorithms were run with a simulation sample size of 50,000. The R code for the implementation of the four simulation algorithms follows.

```
> fitgrid=simcontour(cauchyerrorpost,c(-10,60,1,4.5),difference,
+   50000)
> proposal=list(var=fitlaplace$var,scale=2.5)
> start=array(c(20,3),c(1,2))
> fitrw=rwmetrop(cauchyerrorpost,proposal,start,50000,
+   difference)
> proposal2=list(var=fitlaplace$var,mu=t(fitlaplace$mode))
> fitindep=indepmetrop(cauchyerrorpost,proposal2,start,50000,
```

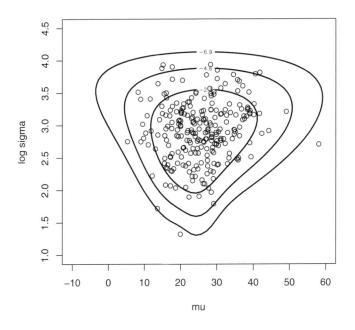

Fig. 6.8. Contour plot of posterior of μ and $\log \sigma$ with simulated sample for Cauchy error model problem.

```
+   difference)
> fitgibbs=gibbs(cauchyerrorpost,start,50000,c(12,.75),
+   difference)
```

The simulated draws for a parameter can be summarized by the computation of the 5th, 50th, and 95th percentiles. For example, one can find the summaries of μ and $\log \sigma$ from the random walk simulation by the command

```
> apply(fitrw$par,2,mean)
```

```
[1] 25.562859  2.843484
```

```
> apply(fitrw$par,2,sd)
```

```
[1] 7.175004 0.372534
```

Table 6.2 displays the estimated posterior quantiles for all of the algorithms described in this chapter. In addition, the acceptance rates for the Metropolis-Hastings random walk and independence chains and the Gibbs sampler are shown. Generally there is agreement among the simulation-based methods and these "exact" posterior summaries are different from the quantiles found using

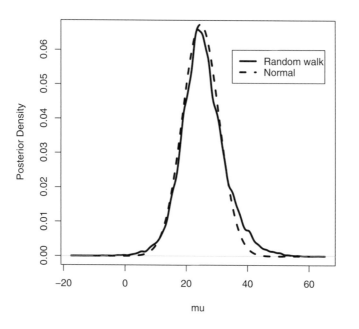

Fig. 6.9. Posterior density of μ using normal approximation and simulated draws from the Metropolis random walk chain.

the Laplace normal approximation. The exact marginal posterior distribution of μ has heavier tails than suggested by the normal approximation; also there is some skewness in the marginal posterior distribution of $\log \sigma$.

Table 6.2. Summaries of the marginal posterior densities of μ and $\log \sigma$ using five computational methods.

Method	Acceptance Rate	μ	$\log \sigma$
Normal approx.		(15.0, 24.7, 34.4)	(2.16, 2.77, 3.38)
Brute force		(14.5, 25.1, 37.7)	(2.22, 2.85, 3.45)
Random walk	.231	(14.8, 25.1, 38.0)	(2.23, 2.85, 3.45)
Independence	.849	(14.4, 25.0, 37.1)	(2.22, 2.85, 3.44)
Gibbs	(.318, .314)	(14.5, 25.2, 38.0)	(2.20, 2.86, 3.45)

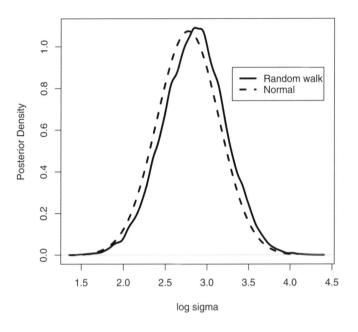

Fig. 6.10. Posterior density of $\log \sigma$ using normal approximation and simulated draws from the Metropolis random walk chain.

6.10 Analysis of the Stanford Heart Transplant Data

Turnbull et al (1974) describe a number of approaches for analyzing heart transplant data from the Stanford Heart Transplanation Program. One of the inferential goals is to decide if heart transplantation extends a patient's life. One of their models, the Pareto model, assumes individual patients in the nontransplant group have exponential lifetime distributions with mean $1/\theta$, where θ is assumed to vary between patients and is drawn from a gamma distribution with density

$$f(\theta) = \frac{\lambda^p}{\Gamma(p)} \theta^{p-1} \exp(-\lambda\theta).$$

Patients in the transplant group have a similar exponential lifetime distribution where the mean is $1/(\theta\tau)$. This model assumes that the patient's risk of death changes by an unknown constant factor $\tau > 0$. If $\tau = 1$, then there is no increased risk by having a transplant operation.

Suppose the survival times $\{x_i\}$ are observed for N nontransplant patients. For n of these patients, x_i represents the actual survival time (in days); the remaining $N-n$ patients were still alive at the end of the study, so x_i represents the censoring time. For the M patients that have a heart transplant, let y_j and z_j denote the time to transplant and survival time; m of these patients died during the study. The unknown parameter vector is (τ, λ, p) with likelihood function given by

$$L(\tau, \lambda, p) = \prod_{i=1}^{n} \frac{p\lambda^p}{(\lambda + x_i)^{p+1}} \prod_{i=n+1}^{N} \left(\frac{\lambda}{\lambda + x_i}\right)^p$$

$$\times \prod_{j=1}^{m} \frac{\tau p\lambda^p}{(\lambda + y_j + \tau z_j)^{p+1}} \prod_{j=m+1}^{M} \left(\frac{\lambda}{\lambda + y_j + \tau z_j}\right)^p,$$

where all the parameters are positive. Suppose we place a uniform prior on (τ, λ, p), and so the posterior density is proportional to the likelihood.

Following our summarization strategy, we transform the parameters by logs:

$$\theta_1 = \log \tau, \quad \theta_2 = \log \lambda, \quad \theta_3 = \log p.$$

The posterior density of $\theta = (\theta_1, \theta_2, \theta_3)$ is given by

$$g(\theta|\text{data}) \propto L(\exp(\theta_1), \exp(\theta_2), \exp(\theta_3)) \prod_{i=1}^{3} \exp(\theta_i).$$

The dataset `stanfordheart` in the LearnBayes package contains the data for 82 patients; for each patient, there are four variables: `survtime`, the survival time; `transplant`, a variable that is 1 or 0 if the patient had a transplant or not; `timetotransplant`, the time a transplant patient waits for the operation; and `state`, a variable that indicates if the survival time was censored (0 if the patient died and 1 if he was still alive). We load this datafile into R.

```
> data(stanfordheart)
```

We write a function `transplantpost` that computes a value of the log posterior. In the following code, we generally follow the earlier notation. The numbers of nontransplant and transplant patients are denoted by N and M. We divide the data into two groups by the transplant indicator variable t. For the nontransplant patients, the survival times and censoring indicators are denoted by xnt and dnt, and for the transplant patients, the waiting times, survival times, and censoring indicators are denoted by y, z, and dt.

```
transplantpost=function(theta,data)
{
x=data[,1]    #  survival time
y=data[,3]    #  time to transplant
t=data[,2]    #  transplant indicator
d=data[,4]    #  censoring indicator (d = 0 if died)
```

```
tau=exp(theta[,1])
lambda=exp(theta[,2])
p=exp(theta[,3])
val=0*tau
xnt=x[t==0]; dnt=d[t==0]
z=x[t==1]; y=y[t==1]; dt=d[t==1]
N=length(xnt)
M=length(z)
for (i in 1:N)
  val=val+(dnt[i]==0)*(p*log(lambda)+log(p)-
          (p+1)*log(lambda+xnt[i]))+
          (dnt[i]==1)*p*log(lambda/(lambda+xnt[i]))
for (i in 1:M)
  val=val+(dt[i]==0)*(p*log(lambda)+log(p*tau)-
          (p+1)*log(lambda+y[i]+tau*z[i]))+
          (dt[i]==1)*p*log(lambda/(lambda+y[i]+tau*z[i]))
val=val+theta[,1]+theta[,2]+theta[,3]
return(val)
}
```

To get an initial idea about the location of the posterior, we run the function laplace. Our initial estimate at the posterior mode is $\theta = (0, 3, -1)$ and we run 10 Newton steps. The algorithm converges and we obtain the posterior mode and an estimate at the variance-covariance matrix.

```
> start=array(c(0,3,-1),c(1,3))
> laplacefit=laplace(transplantpost,start,10,stanfordheart)
> laplacefit

$mode
          [,1]    [,2]       [,3]
[1,] -0.0924209 3.38503 -0.722881

$var
          [,1]          [,2]          [,3]
[1,]  0.17275867 -0.00925073 -0.04994602
[2,] -0.00925073  0.21467648  0.09300626
[3,] -0.04994602  0.09300626  0.06893108

$int
[1] -376.2505
```

We use a Metropolis random walk algorithm (implemented in the function rwmetrop) to simulate from the posterior. We use a proposal variance of $2V$, where V is the estimated variance-covariance matrix from the Laplace fit. We run the simulation for 10,000 iterations and as the output indicates, the acceptance rate was equal to 19%.

```
> proposal=list(var=laplacefit$var,scale=2)
> s=rwmetrop(transplantpost,proposal,start,10000,stanfordheart)
> s$accept
```

[1] 0.1893

One primary inference in this problem is to learn about the three parameters τ, λ, and p. Fig. 6.11 displays density estimates of the simulated draws from the marginal posterior densities of each parameter. These are simply obtained by exponentiating the simulated draws of θ that are output from the function rwmetrop. For example, the first plot in Fig. 6.11 is constructed by first computing the simulated draws of τ and then using the plot(density()) command.

```
> tau=exp(s$par[,1])
> plot(density(tau),main="TAU")
```

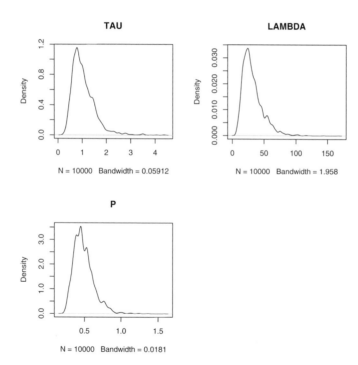

Fig. 6.11. Posterior densities of parameters τ, λ, and p in Pareto survival model.

We can summarize the parameters τ, λ, and p by computing the 5th, 50th, and 95th percentiles of the simulated draws by the apply command.

```
> apply(exp(s$par),2,quantile,c(.05,.5,.95))
```

```
         [,1]      [,2]       [,3]
5%  0.4720614 13.35309 0.3133939
50% 0.9562069 29.01064 0.4746410
95% 2.0703049 63.54526 0.7623879
```

From Fig. 6.11 and these summaries, we see that the value $\tau = 1$ is in the center of the posterior distribution and so there is insufficient evidence to conclude from this data that $\tau \neq 1$. This means that there is insufficient evidence to conclude that the risk of death is higher (or lower) with a transplant operation.

In this problem, one is typically interested in estimating a patient's survival curve. For a nontransplant patient, the survival function is equal to

$$S(t) = \frac{\lambda^p}{(\lambda + t)^p}, \quad t > 0.$$

For a given value of the time t_0, one can compute a sample from the posterior distribution of $S(t_0)$ by computing the function $\lambda^p/(\lambda+t_0)^p$ from the simulated values from the joint posterior distribution of λ and p. In the following code, we assume that simulated samples from the marginal posterior distributions of λ and p are stored in the vectors lambda and p, respectively. Then we (1) set up a grid of values of t and storage vectors p5, p50, and p95; (2) simulate a sample of values of $S(t)$ for each value of t on the grid; and (3) summarize the posterior sample by the computation of the 5th, 50th, and 95th percentiles. These percentiles are stored in the variables p5, p50, and p95. In Fig. 6.12, we graph these percentiles as a function of the time variable t. Since there is little evidence that $\tau \neq 1$, this survival curve represents the risk for both transplant and nontransplant patients.

```
> lambda=exp(s$par[,2])
> t=seq(1,240)
> p5=0*t; p50=0*t; p95=0*t
> for (j in 1:240)
+ { S=(lambda/(lambda+t[j]))^p
+   q=quantile(S,c(.05,.5,.95))
+   p5[j]=q[1]; p50[j]=q[2]; p95[j]=q[3]}
> plot(t,p50,type="l",ylim=c(0,1),ylab="Prob(Survival)",
+   xlab="time")
> lines(t,p5,lty=2)
> lines(t,p95,lty=2)
```

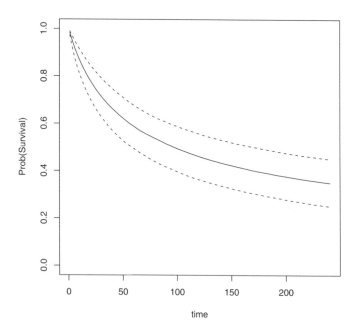

Fig. 6.12. Posterior distribution of probability of survival $S(t)$ for heart transplant patients. Lines correspond to the 5th, 50th, and 95th percentiles of the posterior of $S(t)$ for each time t.

6.11 Further Reading

A good overview of discrete Markov chains is contained in Kemeny and Snell (1976). Since MCMC algorithms currently play a central role in applied Bayesian inference, most modern textbooks devote significant content to these methods. Chapter 11 of Gelman et al (2003) and chapter 5 of Carlin and Louis (2000) provide good introductions to MCMC methods and their application in Bayesian methods. Robert and Casella (2004) and Givens and Hoeting (2005) give more detailed descriptions of MCMC algorithms within the context of computational statistical methods. Introductory discussions of Metropolis and Gibbs sampling are provided, respectively, in Chib and Greenberg (1995) and Casella and George (1992).

6.12 Summary of R Functions

cauchyerrorpost – computes the log posterior density of (M,log S) when a sample is taken from a Cauchy density with location M and scale S and a uniform prior distribution is taken on (M, log S)
Usage: cauchyerrorpost(theta, data)
Arguments: theta, matrix of parameter values where each row represents a value of (M, log S); data, vector containing sample of observations
Value: vector of values of the log posterior where each value corresponds to each row of the parameters in theta

gibbs – implements a Metropolis within Gibbs algorithm for an arbitrary real-valued posterior density defined by the user
Usage: gibbs(logpost,start,m,scale,data)
Arguments: logpost, function defining the log posterior density; start, array with a single row that gives the starting value of the parameter vector; m, the number of iterations of the Gibbs sampling algorithm; scale, vector of scale parameters for the random walk Metropolis steps; data, data used in the function logpost
Value: par, a matrix of simulated values where each row corresponds to a value of the vector parameter; accept, vector of acceptance rates of the Metropolis steps of the algorithm

groupeddatapost – computes the log posterior for (M, log S), when sampling from a normal density and the data are recorded in grouped format
Usage: groupeddatapost=function(theta,data)
Arguments: theta, matrix of parameter values where each row represents a value of (M, log S); data, list with components b, a vector of midpoints, and f, the corresponding bin frequencies
Value: vector of values of the log posterior where each value corresponds to each row of the parameters in theta

indepmetrop – simulates iterates of a Metropolis independence chain for an arbitrary real-valued posterior density defined by the user
Usage: indepmetrop(logpost,proposal,start,m,data)
Arguments: logpost, function defining the log posterior density; proposal, a list containing mu, an estimated mean and var, an estimated variance-covariance matrix of the normal proposal density; start, array with a single row that gives the starting value of the parameter vector; m, the number of iterations of the chain data, data used in the function logpost
Value: par, a matrix of simulated values where each row corresponds to a value of the vector parameter; accept, the acceptance rate of the algorithm.

lbinorm – computes the logarithm of a bivariate normal density
Usage: lbinorm(xy,par)
Arguments: xy, matrix of values where each row corresponds to a value of (x, y); par, list containing m, a vector of means, and v, a variance-covariance matrix

Value: vector of values of the kernel of the log density function

`rwmetrop` – simulates iterates of a random walk Metropolis chain for an arbitrary real-valued posterior density defined by the user
Usage: `rwmetrop(logpost,proposal,start,m,par)`
Arguments: `logpost`, function defining the log posterior density; `proposal`, a list containing `var`, an estimated variance-covariance matrix, and `scale`, the Metropolis scale factor; `start`, array with a single row that gives the starting value of the parameter vector; `m`, the number of iterations of the chain; `par`, data used in the function logpost
Value: `par`, a matrix of simulated values where each row corresponds to a value of the vector parameter; `accept`, the acceptance rate of the algorithm

`transplantpost` – computes the log posterior for (log tau, log lambda, log p) for a Pareto model for survival data
Usage: `transplantpost=function(theta,data)`
Arguments: `theta`, matrix of parameter values where each row represents a value of (log tau, log lambda, log p); `data`, data matrix where columns are survival time, time to transplant, transplant indicator, and censoring indicator
Value: vector of values of the log posterior where each value corresponds to each row of the parameters in theta

6.13 Exercises

1. **A random walk**
   The following matrix represents the transition matrix for a random walk on the integers $\{1, 2, 3, 4, 5\}$.

   $$T = \begin{bmatrix} .2 & .8 & 0 & 0 & 0 \\ .2 & .2 & .6 & 0 & 0 \\ 0 & .4 & .2 & .4 & 0 \\ 0 & 0 & .6 & .2 & .2 \\ 0 & 0 & 0 & .8 & .2 \end{bmatrix}$$

   a) Suppose one starts at the location 1. By use of the `sample` command, simulate 1000 steps of the Markov chain using the probabilities given in the transition matrix. Store the locations of the walk in a vector.
   b) Compute the relative frequencies of the walker in the five states from the simulation output. Guess at the value of the stationary distribution vector w.
   c) Confirm that your guess is indeed the stationary distribution by the matrix computation w %*% T.

2. **Estimating a log-odds with a normal prior**
   In Exercise 1 of Chapter 5, we considered the estimation of a log-odds parameter when y is binomial(n, p) and the log-odds $\theta = \log(p/(1-p))$

is distributed $N(\mu, \sigma)$ with $\mu = 0$ and $\sigma = .25$. The coin was tossed $n = 5$ times and $y = 5$ heads were observed.

Use a Metropolis-Hastings random walk algorithm to simulate from the posterior density. In the algorithm, let s be equal to twice the approximate posterior standard deviation found in the normal approximation. Use the simulation output to approximate the posterior mean and standard deviation of θ, and the posterior probability that θ is positive. Compare your answers with those obtained by the normal approximation in Exercise 1 of Chapter 5.

3. **Genetic linkage model from Rao (2002)**

   In Exercise 2 of Chapter 5, we considered the estimation of a parameter θ in a genetic linkage model. The posterior density was expressed in terms of the real-valued logit $\eta = \log(\theta/(1 - \theta))$.

   a) Use a Metropolis-Hastings random walk algorithm to simulate from the posterior density of η. (Choose the scale parameter s to be twice the approximate posterior standard deviation of η found in a normal approximation.) Compare the histogram of the simulated output of η with the normal approximation. From the simulation output, find a 95% interval estimate for the parameter of interest θ.

   b) Use a Metropolis-Hastings independence algorithm to simulate from the posterior density of η. Use a normal proposal density. Again compare the histogram of the simulated output with the normal approximation and find a 95% probability interval for the parameter of interest θ.

4. **Modeling data with Cauchy errors**

   As in Section 6.8, suppose we observe $y_1, ..., y_n$ from a Cauchy density with location μ and scale σ and a noninformative prior is placed on (μ, σ). Consider the following hypothetical test scores from a class that is a mixture of good and poor students.

36	13	23	6	20	12	23	93
98	91	89	100	90	95	90	87

   The function `cauchyerrorpost` computes the log of the posterior density. A contour plot of the posterior $(\mu, \log \sigma)$ for this data is shown in Fig. 6.13.

   a) Use the `laplace` function to find the posterior mode. Check that you have indeed found the posterior mode by trying several starting values in Newton's algorithm.

   b) Use the Metropolis random walk algorithm (using the function `rwmetrop`) to simulate 1000 draws from the posterior density. Compute the posterior mean and standard deviation of μ and $\log \sigma$.

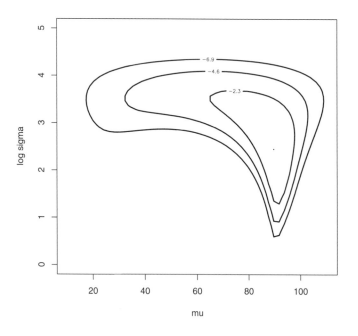

Fig. 6.13. Posterior distribution of μ and $\log \sigma$ for the Cauchy sampling exercise.

5. **Estimation for the two-parameter exponential distribution**
   Exercise 3 of Chapter 5 considered the "type I/time-truncated" life testing experiment. We are interested in the posterior density of $\theta = (\theta_1, \theta_2)$, where $\theta_1 = \log \beta, \theta_2 = \log(t_1 - \mu)$.
   a) Using the posterior mode and variance-covariance matrix from `laplace`, simulate 1000 values from the posterior distribution by the Metropolis random walk algorithm (function `rwmetrop`).
   b) Suppose one is interested in estimating the reliability at time t_0 defined by
   $$R(t_0) = e^{-(t_0 - \mu)/\beta}.$$
   Using your simulated values from the posterior, find the posterior mean and posterior standard deviation of $R(t_0)$ when $t_0 = 10^6$ cycles.

6. **Poisson regression**
   Exercise 4 of Chapter 5 describes an experiment from Haberman (1978) involving subjects reporting one stressful event. The number of events recalled i months before an interview y_i is distributed Poisson with mean λ_i, where the $\{\lambda_i\}$ satisfy the loglinear regression model
   $$\log \lambda_i = \beta_0 + \beta_1 i.$$

One is interested in learning about the posterior density of the regression coefficients (β_0, β_1).

a) Using the output of laplace, construct a Metropolis random walk algorithm for simulating from the posterior density. Use the function rwmetrop to simulate 1000 iterates and compute the posterior mean and standard deviation of β_1.

b) Construct a Metropolis independence algorithm and use the function rwindep to simulate 1000 iterates from the posterior. Compute the posterior mean and standard deviation of β_1.

c) Use a table such as Table 6.2 to compare the posterior estimates using the three computational methods.

7. **Generalized logit model**
Carlin and Louis (2000) describe the use of a generalized logit model to fit dose-mortality data from Bliss (1935). Table 6.3 records the number of adult flour beetles killed after five hours of exposure to various levels of gaseous carbon disulphide. The number of insects killed y_i under dose w_i is assumed binomial(n_i, p_i), where the probability p_i of death is given by

$$p_i = \left(\frac{\exp(x_i)}{1 + \exp(x_i)} \right)^{m_1},$$

where $x_i = (w_i - \mu)/\sigma$. The prior distributions for μ, σ, m_1 are assumed independent, where μ is assigned a uniform prior, σ is assigned a prior proportional to $1/\sigma$, and m_1 is gamma with parameters a_0 and b_0. In the example, the prior hyperparameters of $a_0 = .25$ and $b_0 = 4$ were used. If one transforms to the real-valued parameters $(\theta_1, \theta_2, \theta_3) = (\mu, \log \sigma, \log m_1)$, then Carlin and Louis (2000) show the posterior density is given by

$$g(\theta|\text{data}) \propto \prod_{i=1}^{8} \left[p_i^{y_i} (1 - p_i)^{n_i - y_i} \right] \exp(a_0 \theta_3 - e^{\theta_3}/b_0).$$

Table 6.3. Flour beetle mortality data

Dosage w_i	Number Killed y_i	Number Exposed n_i
1.6907	6	59
1.7242	13	60
1.7552	18	62
1.7842	28	56
1.8113	52	63
1.8369	53	59
1.8610	61	62
1.8839	60	60

a) Write an R function that defines the log posterior of $(\theta_1, \theta_2, \theta_3)$.
b) Carlin and Louis (2000) suggest running a Metropolis random walk chain with a multivariate normal proposal density where the variance-covariance matrix is diagonal with elements 0.00012, 0.033, and 0.10. Use the function rwmetrop to run this chain for 10,000 iterations. Compute the acceptance rate and the 5th and 95th percentiles for each parameter.
c) Run the function laplace to get a non-diagonal estimate of the variance-covariance matrix. Use this estimate in the proposal density of rwmetrop and run the chain for 10,000 iterations. Compute the acceptance rate and the 5th and 95th percentiles for each parameter.
d) Compare your answers in parts (b) and (c).

7

Hierarchical Modeling

7.1 Introduction

In this chapter, we illustrate the use of R to summarize an exchangeable hierarchical model. We begin by giving a brief introduction to hierarchical modeling. Then we consider the simultaneous estimation of the true mortality rates from heart transplants for a large number of hospitals. Some of the individual estimated mortality rates are based on limited data and it may be desirable to combine the individual rates in some way to obtain more accurate estimates. We describe a two-stage model, a mixture of gamma distributions, to represent prior beliefs that the true mortality rates are exchangeable. We describe the use of R to simulate from the posterior distribution. We first use contour graphs and simulation to learn about the posterior distribution of the hyperparameters. Once we simulate hyperparameters, we can simulate from the posterior distributions of the true mortality rates from gamma distributions. We conclude by illustrating how the simulation of the joint posterior can be used to perform different types of inferences in the heart transplant application.

7.2 Introduction to Hierarchical Modeling

In many statistical problems, we are interested in learning about many parameters that are connected in some way. To illustrate, consider the following three problems described in this chapter and the chapters to follow.

1. **Simultaneous estimation of hospital mortality rates**

   In the main example of this chapter, one is interested in learning about the mortality rates due to heart transplant surgery for 94 hospitals. Each hospital has a true mortality rate λ_i, and so one wishes to simultaneously estimate the 94 rates $\lambda_1, ..., \lambda_{94}$. It is reasonable to believe a priori that the true rates are similar in size, which implies a dependence structure between the parameters. If one is told some information about a particular hospital's true rate, that information would likely affect one's belief about the location of a second hospital's rate.

2. **Estimating college grade point averages**

   In an example in Chapter 10, admissions people at a particular university collect a table of means of freshman grade point averages (GPA) organized by the student's high school rank and his or her score on a standardized test. One wishes to learn about the collection of population mean GPAs with the ultimate goal of making predictions about the success of future students that attend the university. One believes that the population GPAs can be represented as a simple linear function of the high school rank and standardized test score.

3. **Estimating career trajectories**

   In an example in Chapter 11, one is learning about the pattern of performance of athletes as they age during their sports careers. In particular, one wishes to estimate the *career trajectories* of the batting performances of a number of baseball players. For each player, one fits a model to estimate his career trajectory, and Fig. 7.1 displays the fitted career trajectories for nine players. Note that the shapes of these trajectories are similar; a player generally will increase in performance until his late 20s or early 30s and then decline until retirement. The prior belief is that the true trajectories will be similar between players, which again implies a prior distribution with dependence.

In many-parameter situations like the ones described here, it is natural to construct a prior distribution in a *hierarchical* fashion. In this type of model, the observations are given distributions conditional on parameters, and the parameters in turn have distributions conditional on additional parameters called hyperparameters. Specifically, we begin by specifying a data distribution

$$y \sim f(y|\theta),$$

and the prior vector θ will be assigned a prior distribution with unknown hyperparameters λ:

$$\theta \sim g_1(\theta|\lambda).$$

The hyperparameter vector λ in turn will be assigned a distribution

$$\lambda \sim g_2(\lambda).$$

One general way of constructing a hierarchical prior is based on the prior belief of *exchangeability*. A set of parameters $\theta = (\theta_1, ..., \theta_k)$ is exchangeable if the distribution of θ is unchanged if the parameter components are permuted. This implies that one's prior belief about θ_j, say, will be the same as one's belief about θ_h. One can construct an exchangeable prior by assuming that the components of θ are a random sample from a distribution g_1:

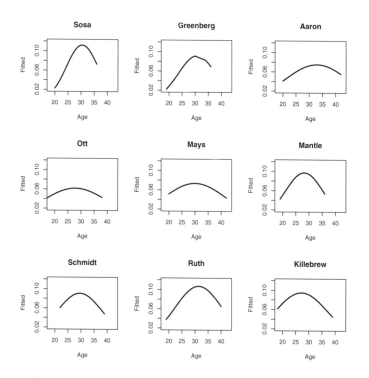

Fig. 7.1. Plots of fitted career trajectories for nine baseball players as a function of their age.

$$\theta_1, ..., \theta_k \text{ random sample from } g_1(\theta|\lambda),$$

and the unknown hyperparameter vector λ is assigned a known prior at the second stage:

$$\lambda \sim g_2(\lambda).$$

This particular form of hierarchical prior will be used for the mortality rates example of this chapter and the career trajectories example of Chapter 11.

7.3 Individual and Combined Estimates

Consider again the heart transplant mortality data discussed in Chapter 3. The number of deaths within 30 days of heart transplant surgery is recorded for each of 94 hospitals. In addition, we record for each hospital an expected number of deaths called the exposure denoted by e. We let y_i and e_i denote the respective observed number of deaths and exposure for the ith hospital. In R, we read in the relevant dataset **hearttransplants** in the LearnBayes package.

```
> data(hearttransplants)
> attach(hearttransplants)
```

A standard model assumes that the number of deaths y_i follows a Poisson distribution with mean $e_i \lambda_i$ and the objective is to estimate the mortality rate per unit exposure λ_i. The fraction y_i/e_i is the number of deaths per unit exposure and can be viewed as an estimate of the death rate for the ith hospital. In Fig. 7.2, we plot the ratios $\{y_i/e_i\}$ against the logarithms of the exposures $\{\log(e_i)\}$ for all hospitals where each point is labeled by the number of observed deaths y_i.

```
> plot(log(e), y/e, pch = as.character(y))
```

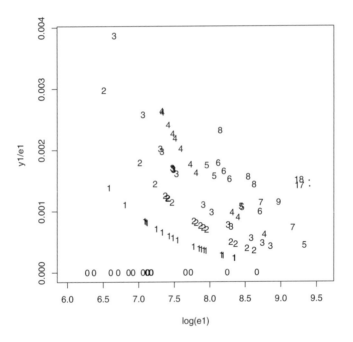

Fig. 7.2. Plot of death rates against log exposure for all hospitals. Each point is labeled by the number of observed deaths.

Note that the estimated rates are highly variable, especially for programs with small exposures. The programs experiencing no deaths (a plotting label of 0) also are primarily associated with small exposures.

Suppose we are interested in simultaneously estimating the true mortality rates $\{\lambda_i\}$ for all hospitals. One option is to simply estimate the true rates by

the individual death rates
$$\frac{y_1}{e_1}, \ldots, \frac{y_{94}}{e_{94}}.$$

Unfortunately these individual rates can be poor estimates, especially for the hospitals with small exposures. In Fig. 7.2, we saw that some of these hospitals did not experience any deaths and the individual death rate $y_i/e_i = 0$ would likely underestimate the hospital's true mortality rate. Also it is clear from the figure that the rates for the hospitals with small exposures have high variability.

Since the individual death rates can be poor, it seems desirable to combine the individual estimates in some way to obtain improved estimates. Suppose we can assume that the true mortality rates are equal across hospitals; that is,

$$\lambda_1 = \ldots = \lambda_{94}.$$

Under this "equal-means" Poisson model, the estimate of the mortality rate for the ith hospital would be the pooled estimate

$$\frac{\sum_{j=1}^{94} y_j}{\sum_{j=1}^{94} e_j}.$$

But this pooled estimate is based on the strong assumption that the true mortality rate is the same across hospitals. This is questionable since one would expect some variation in the true rates.

We have discussed two possible estimates for the mortality rate of the ith hospital: the individual estimate y_i/e_i and the pooled estimate $\sum y_j / \sum e_j$. A third possibility is the compromise estimate

$$(1-\lambda)\frac{y_i}{e_i} + \lambda \frac{\sum_{j=1}^{94} y_j}{\sum_{j=1}^{94} e_j}.$$

This estimate shrinks or moves the individual estimate y_i/e_i toward the pooled estimate $\sum y_j / \sum e_j$ where the parameter $0 < \lambda < 1$ determines the size of the shrinkage. We will see that this shrinkage estimate is a natural byproduct of the application of an exchangeable prior model on the true mortality rates.

7.4 Equal Mortality Rates?

Before we consider an exchangeable model, let's illustrate fitting and checking the model where the mortality rates are assumed equal. Suppose y_i is distributed Poisson$(e_i \lambda)$, $i = 1, \ldots, 94$, and the common mortality rate λ is assigned a standard noninformative prior of the form

$$g(\lambda) \propto \frac{1}{\lambda}.$$

Then the posterior density of λ is given by

$$g(\lambda|\text{data}) \propto \frac{1}{\lambda} \prod_{j=1}^{94} \left[\lambda^{y_j} \exp(-e_j\lambda) \right]$$

$$= \lambda^{\sum_{j=1}^{94} y_j - 1} \exp(-\sum_{j=1}^{94} e_j\lambda)$$

which is recognized as a gamma density with parameters $\sum_{j=1}^{94} y_j$ and $\sum_{j=1}^{94} e_j$. For our data, we compute

```
> sum(y)
```

```
[1] 277
```

```
> sum(e)
```

```
[1] 294681
```

and so the posterior density for the common rate λ is gamma(277, 294681).

One general Bayesian method of checking the suitability of a fitted model such as this is based on the posterior predictive distribution. Let y_i^* denote the number of transplant deaths for hospital i with exposure e_i in a future sample. Conditional on the true rate λ, y_i^* has a Poisson distribution with mean $e_i\lambda$. Our current beliefs about the ith true rate are contained in the posterior density $g(\lambda|y)$. The unconditional distribution of y_i^*, the posterior predictive density, is given by

$$f(y_i^*|e_i, y) = \int f_P(y_i^*|e_i\lambda)g(\lambda|y)d\lambda,$$

where $f_P(y|\lambda)$ is the Poisson sampling density with mean λ. The posterior predictive density represents the likelihood of future observations based on our fitted model. For example, the density $f(y_i^*|e_i, y)$ represents the number of transplants that we would predict in the future for a hospital with exposure e_i. If the actual number of observed deaths y_i is in the middle of this predictive distribution, then we can say that our observation is consistent with our model fit. On the other hand, if the observed y_i is in the extreme tails of the distribution $f(y_i^*|e_i, y)$, then this observation indicates that the model is inadequate in fitting this observation.

To illustrate the use of the posterior predictive distribution, consider hospital 94 that had 17 transplant deaths, that is, $y_{94} = 17$. Did this hospital have an unusually high number of deaths? To answer this question, we simulate 1000 values from the posterior predictive density of y_{94}^*.

To simulate from the predictive distribution of y_{94}^*, we first simulate 1000 draws of the posterior density of λ

```
> lambda=rgamma(1000,shape=277,rate=294681)
```

and then simulate draws of y_{94}^{*} from a Poisson distribution with mean $e_{94}\lambda$.

```
> ys94=rpois(1000,e[94]*lam)
```

Using the following R code, Fig. 7.3 displays a histogram of this posterior predictive distribution and the actual number of transplant deaths y_{94} is shown by a vertical line.

```
> hist(ys94,breaks=seq(1.5,26.5,by=1))
> lines(c(y[94],y[94]),c(0,120),lwd=3)
```

Since the observed y_j is in the tail portion of the distribution, it seems inconsistent with the fitted model – it suggests that this hospital actually has a higher true mortality rate than estimated from this equal-rates model.

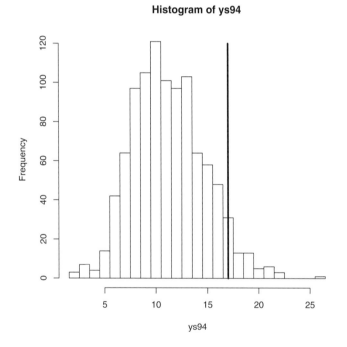

Fig. 7.3. Histogram of simulated draws from the posterior predictive distribution of y_{94}^{*}. The actual number of transplant deaths is shown by a vertical line.

We can check the consistency of the observed y_i with its posterior predictive distribution for all hospitals. For each distribution, we compute the probability that the future observation y_i^{*} is at least as extreme as y_i:

$$\min\{P(y_i^* \leq y_i), P(y_i^* \geq y_i)\}.$$

The following R code computes the probabilities of "at least as extreme" for all observations and places the probabilities in the vector pout.

```
> pout=0*y
> lambda=rgamma(1000,shape=277,rate=294681)
> for (i in 1:94){
+    ysi=rpois(1000,e[i]*lambda)
+    pleft=sum(ysi<=y[i])/1000
+    pright=sum(ysi>=y[i])/1000
+    pout[i]=min(pleft,pright)
+ }
```

We plot the probabilities against the log exposures which is displayed in Fig. 7.4.

```
> plot(log(e),pout,ylab="Prob(extreme)")
```

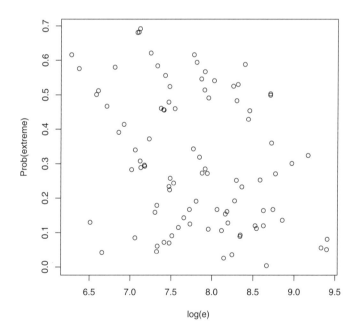

Fig. 7.4. Scatterplot of predictive probabilities of "at least as extreme" against log exposures for all observations.

Note that a number of these tail probabilities appear small (15 are smaller than 0.10) which means that the "equal rates" model is inadequate for explaining the distribution of mortality rates for the group of 94 hospitals. We will have to assume differences between the true mortality rates that will be modeled by the exchangeable model described in the next section.

7.5 Modeling a Prior Belief of Exchangeability

At the first stage of the prior, the true death rates $\lambda_1, ..., \lambda_{94}$ are assumed to be a random sample from a gamma$(\alpha, \alpha/\mu)$ distribution of the form

$$g(\lambda|\alpha, \mu) = \frac{(\alpha/\mu)^\alpha \lambda^{\alpha-1} \exp(-\alpha\lambda/\mu)}{\Gamma(\alpha)}, \lambda > 0.$$

The prior mean and variance of λ are given by μ and μ^2/α, respectively. At the second stage of the prior, the hyperparameters μ and α are assumed independent, with μ assigned a gamma(a, b) distribution with density $\mu^{a-1} \exp(-b\mu)$ and α the density $g(\alpha)$.

This prior distribution induces positive correlation between the true death rates. To see this, suppose one assigns the hyperparameter μ a gamma(10, 10) distribution and sets the hyperparameter α equal to a fixed value α_0. (This is equivalent to assigning a density $g(\alpha)$ that places probability one on the value α_0.) One can simulate values of, say (λ_1, λ_2), from the prior distribution by

- simulating values from μ from the gamma(a, b) distribution, α from the prior density $g(\alpha)$
- for each simulated pair (μ, α), simulate λ_1, λ_2 from gamma$(\alpha, \alpha/\mu)$ distributions

This simulation is illustrated in the following R code. Fig. 7.5 displays 500 simulated values from the prior distribution of (λ_1, λ_2) for the values α_0 equal to 5, 20, 80, and 400. Note that since μ is assigned a gamma(10, 10) distribution, both the true rates λ_1 and λ_2 are centered about the value 1. The hyperparameter α is a precision parameter that controls the correlation between the parameters. For the fixed value $\alpha = 400$, note that λ_1 and λ_2 are concentrated along the line $\lambda_1 = \lambda_2$. As the precision parameter α approaches infinity, the exchangeable prior places all of its mass along the space where $\lambda_1 = ... = \lambda_{94}$.

```
> par(mfrow = c(2, 2))
> m = 500
> alphas = c(5, 20, 80, 400)
> for (j in 1:4) {
+      mu = rgamma(m, shape = 10, rate = 10)
+      lambda1 = rgamma(m, shape=alphas[j], rate=alphas[j]/mu)
+      lambda2 = rgamma(m, shape=alphas[j], rate=alphas[j]/mu)
```

```
+      plot(lambda1, lambda2)
+      title(main=paste("alpha=",as.character(alphas[j])))
+ }
```

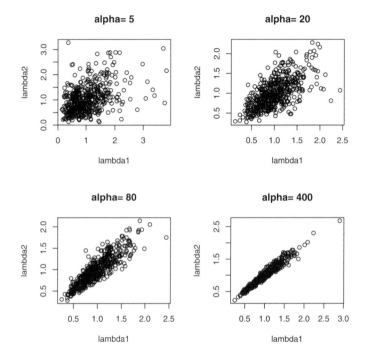

Fig. 7.5. Simulated values from the exchangeable prior on (λ_1, λ_2) for values of the precision parameter $\alpha = 5, 20, 80,$ and 400.

Although we used subjective priors to illustrate the behavior of the prior distribution, in practice vague distributions can be chosen for the hyperparameters μ and α. In this example, we assign the mean parameter the typical vague prior of the form

$$g(\mu) \propto \frac{1}{\mu}, \mu > 0.$$

The precision parameter α assigned the proper, but relatively flat, prior density of the form

$$g(\alpha) = \frac{z_0}{(\alpha + z_0)^2}, \alpha > 0.$$

The user will specify a value of the parameter z_0 that is the median of α. In this example, we let $z_0 = 0.53$.

7.6 Posterior Distribution

Owing to the conditionally independent structure of the hierarchical model and the choice of a conjugate prior form at stage 2, there is a relatively simple posterior analysis. Conditional on values of the hyperparameters μ and α, the rates $\lambda_1, ..., \lambda_{94}$ have independent posterior distributions. The posterior distribution of λ_i is gamma$(y_i + \alpha, e_i + \alpha/\mu)$. The posterior mean of λ_i, conditional on α and μ, can be written as

$$E(\lambda_i | y, \alpha, \mu) = \frac{y_i + \alpha}{e_i + \alpha/\mu} = (1 - B_i)\frac{y_i}{e_i} + B_i\mu,$$

where

$$B_i = \frac{\alpha}{\alpha + e_i\mu}.$$

The posterior mean of the true rate λ_i can be viewed as a shrinkage estimator, where B_i is the shrinkage fraction of the posterior mean away from the usual estimate y_i/e_i toward the prior mean μ.

Also since a conjugate model structure was used, the rates λ_i can be integrated out of the joint posterior density, resulting in the marginal posterior density of (α, μ):

$$p(\alpha, \mu | \text{data}) = K\frac{1}{\Gamma^{94}(\alpha)} \prod_{j=1}^{94} \left[\frac{(\alpha/\mu)^\alpha \Gamma(\alpha + y_i)}{(\alpha/\mu + e_i)^{(\alpha+y_i)}}\right] \frac{z_0}{(\alpha + z_0)^2} \frac{1}{\mu},$$

where K is a proportionality constant.

7.7 Simulating from the Posterior

In the previous section the posterior density of all parameters was expressed as

$$g(\text{hyperparameters}|\text{data}) \, g(\text{true rates}|\text{hyperparameters, data}),$$

where the hyperparameters are (μ, α) and the true rates are $(\lambda_1, ..., \lambda_{94})$. By the composition method, we can simulate a random draw from the joint posterior by

- simulating (μ, α) from the marginal posterior distribution
- simulating $\lambda_1, ..., \lambda_{94}$ from their distribution conditional on the values of the simulated μ and α

First we need to simulate from the marginal density of the hyperparameters μ and α. Since both parameters are positive, a good first step in this simulation process is to transform each to the real-valued parameters

$$\theta_1 = \log(\alpha), \theta_2 = \log(\mu).$$

The marginal posterior of the transformed parameters is given by

$$p(\theta_1, \theta_2|\text{data}) = K \frac{1}{\Gamma^{94}(\alpha)} \prod_{j=1}^{94} \left[\frac{(\alpha/\mu)^\alpha \Gamma(\alpha + y_i)}{(\alpha/\mu + e_i)^{(\alpha+y_i)}} \right] \frac{z_0 \alpha}{(\alpha + z_0)^2}.$$

The following R function `poissgamexch` contains the definition of the log posterior of θ_1 and θ_2.

```
poissgamexch=function(theta,datapar)
{
y=datapar$data[,2]; e=datapar$data[,1]
z0=datapar$z0
alpha=exp(theta[,1]); mu=exp(theta[,2])
beta=alpha/mu
N=length(y)
val=0*alpha;
for (i in 1:N)
{
val=val+lgamma(alpha+y[i])-(y[i]+alpha)*log(e[i]+beta)+
  alpha*log(beta)
}
val=val-N*lgamma(alpha)+log(alpha)-2*log(alpha+z0)
return(val)
}
```

Note that this function has two inputs:

- `theta`– a matrix of two columns where each row corresponds to a value of (θ_1, θ_2)
- `datapar` – a R list with two components, the `data` and the value of the hyperparameter `z0`

Note that since `theta` is a matrix, we sum over the observations to compute the log posterior. We use the function `lgamma` that computes the log of the gamma function, $\log \Gamma(x)$.

Using the R function `laplace`, we find the posterior mode and associated variance-covariance matrix. We perform five iterations of the Newton-Raphson algorithm at the starting value $(\theta_1, \theta_2) = (2, -7)$. The output of `laplace` includes the mode and the corresponding estimate at the variance-covariance matrix.

```
> datapar = list(data = hearttransplants, z0 = 0.53)
> start=array(c(2, -7), c(1, 2))
> fit = laplace(poissgamexch, start, 5, datapar)
> fit

$mode
        [,1]        [,2]
```

```
[1,] 1.88535 -6.955614
```

```
$var
            [,1]            [,2]
[1,]   0.23412668 -0.003077430
[2,] -0.00307743  0.005863179
```

```
$int
[1] -2208.502
```

This output gives us information about the location of the posterior density. By trial and error, we use the function `mycontour` to find a grid that contains the posterior density of (θ_1, θ_2). The resulting graph is displayed in Fig. 7.6.

```
> par(mfrow = c(1, 1))
> mycontour(poissgamexch, c(0, 8, -7.3, -6.6), datapar)
> title(xlab="log alpha",ylab="log mu")
```

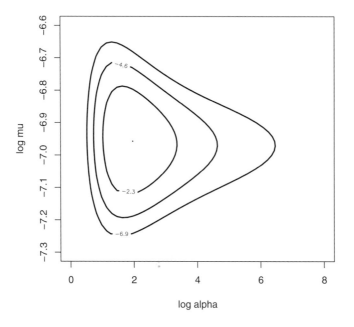

Fig. 7.6. Contour plot of the posterior density of $(\log \alpha, \log \mu)$ for the heart transplant example. Contour lines are drawn at 10%, 1%, and .1% of the modal value.

By inspection of Fig. 7.6, we see that the posterior density for (θ_1, θ_2) is nonnormal shaped, especially in the direction of $\theta_1 = \log \alpha$. Since the normal approximation to the posterior is inadequate, we obtain a simulated sample of (θ_1, θ_2) by use of the "Metropolis within Gibbs" algorithm in the function gibbs. In this Gibbs sampling algorithm, we start at the value $(\theta_1, \theta_2) = (4, -7)$ and iterate through 1000 cycles with Metropolis scale parameters $c_1 = 1, c_2 = .15$. As the output indicates, the acceptance rates in the simulation of the two conditional distributions are each about 30%.

```
> start = array(c(4, -7), c(1, 2))
> fitgibbs = gibbs(poissgamexch, start, 1000, c(1,.15), datapar)
> fitgibbs$accept

      [,1]  [,2]
[1,] 0.312 0.284
```

Fig. 7.7 shows a simulated sample of 1000 placed on top of the contour graph. Note that most of the points fall within the first two contour lines of the graph, indicating that the algorithm appears to give a representative sample from the marginal posterior distribution of θ_1 and θ_2.

```
> mycontour(poissgamexch, c(0, 8, -7.3, -6.6), datapar)
> points(fitgibbs$par[, 1], fitgibbs$par[, 2])
```

Fig. 7.8 shows a kernel density estimate of the simulated draws from the marginal posterior distribution of the precision parameter $\theta_1 = \log(\alpha)$.

```
> plot(density(fitgibbs$par[, 1], bw = 0.2))
```

We can learn about the true mortality rates $\lambda_1, ..., \lambda_{94}$ by simulating values of from their posterior distributions. Given values of the hyperparameters α and μ, the true rates have independent posterior distributions with λ_i distributed gamma$(y_i + \alpha, e_i + \alpha/\mu)$. For each rate, we use the R rgamma function to obtain a sample from the gamma distribution, where the gamma parameters are functions of the simulated values of α and μ. For example, one can obtain a sample from the posterior distribution of λ_1 by the R code

```
> alpha = exp(fitgibbs$par[, 1])
> mu = exp(fitgibbs$par[, 2])
> lam1 = rgamma(1000, y[1] + alpha, e[1] + alpha/mu)
```

After we obtain a simulated sample of size 1000 for each true rate λ_i, we can summarize each sample by computing the 5th and 95th percentiles. The interval from these two percentiles constitutes an approximate 90% probability interval for λ_i. We graph these 90% probability intervals as vertical lines on our original graph of the log exposures and the individual rates in Fig. 7.9. In contrast to the wide variation in the observed death rates, note the similarity in the locations of the probability intervals for the true rates. This indicates

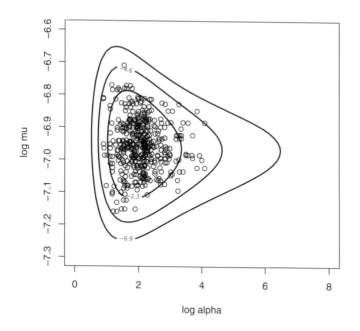

Fig. 7.7. Contour plot of the posterior density of $(\log \alpha, \log \mu)$ for the heart transplant example with a sample of simulated values placed on top.

that these Bayesian estimates are shrinking the individual rates toward the pooled estimate.

```
> alpha = exp(fitgibbs$par[, 1])
> mu = exp(fitgibbs$par[, 2])
> plot(log(e), y/e, pch = as.character(y))
> for (i in 1:94) {
+     lami = rgamma(1000, y[i] + alpha, e[i] + alpha/mu)
+     probint = quantile(lami, c(0.05, 0.95))
+     lines(log(e[i]) * c(1, 1), probint)
+ }
```

7.8 Posterior Inferences

Once a simulated sample of true rates $\{\lambda_i\}$ and the hyperparameters μ, α has been generated from the joint posterior distribution, we can use this sample to perform various types of inferences.

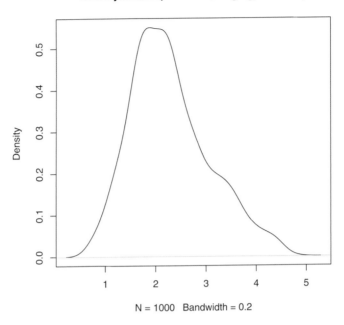

Fig. 7.8. Density estimate of simulated draws from the marginal posterior of $\log \alpha$.

7.8.1 Shrinkage

The posterior mean of the ith true mortality rate λ_i can be approximated by

$$E(\lambda_i|\text{data}) \approx (1 - E(B_i|\text{data}))\frac{y_i}{e_i} + E(B_i|\text{data})\frac{\sum_{j=1}^{94} y_j}{\sum_{j=1}^{94} e_j},$$

where $B_i = \alpha/(\alpha + e_i\mu)$ is the size of the shrinkage of the ith observed rate y_i/e_i toward the pooled estimate $\sum_{j=1}^{94} y_j / \sum_{j=1}^{94} e_j$. In the following R code, we compute the posterior mean of the shrinkage sizes $\{B_i\}$ for all 94 components. In Fig. 7.10, we plot the mean shrinkages against the logarithms of the exposures. For the hospitals with small exposures, the Bayesian estimate shrinks the individual estimate 90% toward the combined estimate. In contrast, for large hospitals with high exposures, the shrinkage size is closer to 50%.

```
> shrinkage = 0 * e
> for (i in 1:94) shrinkage[i] = mean(alpha/(alpha + e[i] * mu))

> plot(log(e), shrinkage)
```

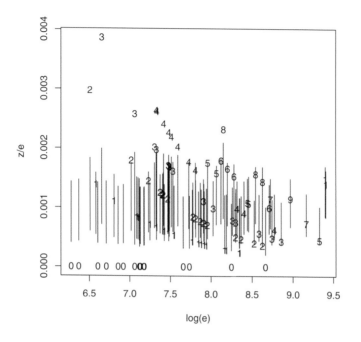

Fig. 7.9. Plot of observed death rates against log exposures together with intervals representing 90% posterior probability bands for the true rates $\{\lambda_i\}$.

7.8.2 Comparing Hospitals

Suppose one is interested in comparing the true mortality rates of the hospitals. Specifically, suppose one wishes to compare the "best hospital" with the other hospitals. First, we find the hospital with the smallest estimated mortality rate. In the following R output, we compute the posterior mean of the mortality rates, where the posterior mean of the true rate for hospital i is given by

$$E\left(\frac{y_i + \alpha}{e_i + \alpha/\mu}\right),$$

where the expectation is taken over the marginal posterior distribution of (α, μ).

```
> hospital=1:94
> meanrate=array(0,c(94,1))
> for (i in 1:94)
+ meanrate[i]=mean(rgamma(1000, y[i] + alpha, e[i] + alpha/mu))
> hospital[meanrate==min(meanrate)]
```

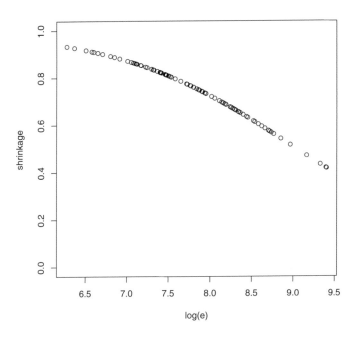

Fig. 7.10. Plot of the posterior shrinkages against the log exposures for the heart transplant example.

```
[1] 85
```

We identify hospital 85 as the one with the smallest true mortality rate.

Suppose we wish to compare hospital i with hospital j. One first obtains simulated draws from the marginal distribution of (λ_i, λ_j). Then the probability that hospital i has a smaller mortality rate, $P(\lambda_i < \lambda_j)$, can be estimated by the proportion of simulated (λ_i, λ_j) pairs where λ_i is smaller than λ_j. In the following R code, we compute these probabilities for all pairs of hospitals and the results are stored in the matrix **better**. The probability that hospital i's rate is smaller than hospital j's rate is stored in the ith row and jth element of **better**.

```
> better=array(0,c(94,94))
> for (i in 1:94){
+    for (j in (i+1):94){
+    if (j <=94) {
+    lami=rgamma(1000,y[i]+alpha,e[i]+alpha/mu)
+    lamj=rgamma(1000,y[j]+alpha,e[j]+alpha/mu)
+    better[i,j]=mean(lami<lamj)
```

```
+    better[j,i]=1-better[i,j]
+ }}}
```

To compare the best hospital 85 with the remaining hospitals, we display the 85th column of the matrix **better**. These give the probabilities $P(\lambda_i < \lambda_{85})$ for all i. We display these probabilities for the first 24 hospitals. Note that hospital 85 is better than most of these hospitals since most of the posterior probabilities are close to zero.

```
> better[1:24,85]
```

```
 [1] 0.166 0.184 0.078 0.114 0.131 0.217 0.205 0.165 0.040 0.196
[11] 0.192 0.168 0.184 0.071 0.062 0.196 0.231 0.056 0.303 0.127
[21] 0.160 0.135 0.041 0.070
```

7.9 Posterior Predictive Model Checking

In Section 7.3, we used the posterior predictive distribution to examine the suitability of the "equal rates" model where $\lambda_1 = \ldots = \lambda_{94}$, and we saw that the model seemed inadequate in explaining the number of transplant deaths for individual hospitals. Here we use the same methodology to check the appropriateness of the exchangeable model.

Again we consider hospital 94, which experienced 17 deaths. Recall that simulated draws of the hyperparameters α and μ are contained in the vectors **alpha** and **mu**, respectively. To simulate from the predictive distribution of y_{94}^* we first simulate draws of the posterior density of λ_{94}

```
> lam94=rgamma(1000,y[94]+alpha,e[94]+alpha/mu)
```

and then simulate draws of y_{94}^* from a Poisson distribution with mean $e_{94}\lambda_{94}$.

```
> ys24=rpois(1000,e[94]*lam94)
```

Fig. 7.11 displays the histogram of y_{94}^* and places a vertical line on top corresponding to the value $y_{94} = 17$ using the commands

```
> hist(ys94,breaks=seq(1.5,39.5,by=1))
> lines(y[94]*c(1,1),c(0,100),lwd=3)
```

Note that in this case the observed number of deaths for this hospital is in the middle of the predictive distribution that indicates agreement of this observation with the fitted model.

We can perform diagnostics for this exchangeable model by checking the consistency of the observed y_i with its posterior predictive distribution for all hospitals. In the following R code, we compute the probability that the future observation y_i^* is at least as extreme as y_i for all observations; the probabilities are placed in the vector **pout.exchange**.

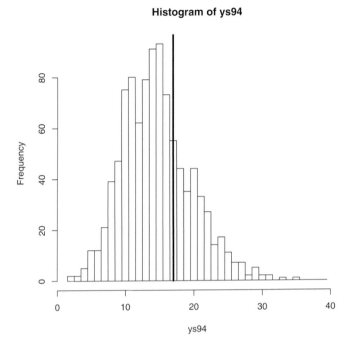

Fig. 7.11. Histogram of posterior predictive distribution of y_{94}^* for hospital 94 from the exchangeable model. The observed value of y_{94} is indicated by the vertical line.

```
> pout.exchange=0*y
> for (i in 1:94){
+    lami=rgamma(1000,y[i]+alpha,e[i]+alpha/mu)
+    ysi=rpois(1000,e[i]*lami)
+    pleft=sum(ysi<=y[i])/1000
+    pright=sum(ysi>=y[i])/1000
+    pout.exchange[i]=min(pleft,pright)
+ }
```

Recall that the probabilities of at least as extreme for the equal means model were contained in the vector pout. To compare the goodness of fits of the two models, Fig. 7.12 shows a scatterplot of the two sets of probabilities with a comparison line $y = x$ placed on top.

```
> plot(pout,pout.exchange,xlab="P(extreme), equal means",
+ ylab="P(extreme), exchangeable")
> abline(0,1)
```

Note that the probabilities of extreme for the exchangeable model are larger, indicating that the observations are more consistent with the exchangeable fitted model. Note that only two of the observations have a probability smaller than 0.1 for the exchangeable model, indicating general agreement of the observed data with this model.

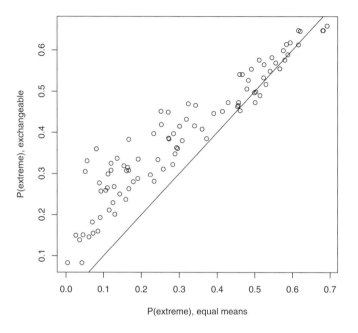

Fig. 7.12. Scatterplot of posterior predictive probabilities of "at least as extreme" for the equal means and exchangeable models.

7.10 Further Reading

Gelman et al (2003), chapter 5, provide a good introduction to hierarchical models. Carlin and Louis (2000), chapter 3, introduce hierarchical modeling from an empirical Bayes perspective. Posterior predictive model checking is described as a general method for model checking in chapter 6 of Gelman et al (2003). The use of hierarchical modeling to analyze the heart transplant data is described in Christiansen and Morris (1995).

7.11 Summary of R Functions

`poissgamexch` – computes the logarithm of the posterior for the parameters (log alpha, log mu) in a Poisson/gamma model
Usage: `poissgamexch(theta,datapar)`
Arguments: `theta`, matrix of parameter values where each row represents a value of (log alpha, log mu); `datapar`, list with components `data` (matrix with column of counts and column of exposures) and `z0`, the value of the second-stage hyperparameter
Value: vector of values of the log posterior where each value corresponds to each row of the parameters in theta

7.12 Exercises

1. **Normal/normal exchangeable model**
   Suppose we have J independent experiments, where in the jth experiment, we observe the single observation y_j that is normally distributed with mean θ_j and known variance σ_j^2. Suppose the parameters $\theta_1, ..., \theta_J$ are drawn from a normal population with mean μ and variance τ^2. The vector of hyperparameters (μ, τ) is assigned a uniform prior. Gelman et al (2003) describe the posterior calculations for this model. To summarize,

   * Conditional on the hyperparameters μ and τ, the θ_j have independent posterior distributions, where $\theta_j | \mu, \tau, y$ is normally distributed with mean $\hat{\theta}_j$ and variance V_j, where

   $$\hat{\theta}_j = \frac{y_j/\sigma_j^2 + \mu/\tau^2}{1/\sigma_j^2 + 1/\tau^2}, \ V_j = \frac{1}{1/\sigma_j^2 + 1/\tau^2}.$$

   * The marginal posterior density of the hyperparameters (μ, τ) is given by

   $$g(\mu, \tau | y) \propto \prod_{j=1}^{J} \phi(y_j | \mu, \sqrt{\sigma_j^2 + \tau^2}),$$

   where $\phi(y|\mu, \sigma)$ denotes the normal density with mean μ and variance σ.

   To illustrate this model, Gelman et al (2003) describe the results of independent experiments to determine the effects of special coaching programs on SAT scores. For the jth experiment, one observes an estimated coaching effect y_j with associated standard error σ_j; the values of the effects and standard errors are displayed in Table 7.1. The objective is to combine the coaching estimates in some way to obtain improved estimates at the true effects θ_j.

Table 7.1. Observed effects of special preparation on SAT scores in eight randomized experiments.

School	Treatment Effect y_j	Standard Error σ_j
A	28	15
B	8	10
C	-3	16
D	7	11
E	-1	9
F	1	11
G	18	10
H	12	18

a) Write an R function to compute the logarithm of the posterior density of the hyperparameters μ and $\log \tau$. (Don't forget to include the Jacobian term in the transformation to $(\mu, \log \tau)$.) Use a simulation algorithm such as Gibbs sampling (function `gibbs`), random walk Metropolis (function `rwmetrop`), or independence Metropolis (function `indepmetrop`) to obtain a sample of size 1000 from the posterior of $(\mu, \log \tau)$.

b) Using the simulated sample from the marginal posterior of $(\mu, \log \tau)$, simulate 1000 draws from the joint posterior density of the means $\theta_1, ..., \theta_J$. Summarize the posterior distribution of each θ_j by the computation of a posterior mean and posterior standard deviation.

2. **Normal/normal exchangeable model (continued)**

   We assume that the sampling algorithm in Exercise 7.1 has been followed and one has simulated a sample of 1000 values from the marginal posterior of the hyperparameters μ and $\log \tau$, and also from the posterior densities of $\theta_1, ..., \theta_J$.

   a) The posterior mean of θ_j, conditional on μ and τ, can be written as

   $$E(\theta_j|y, \mu, \tau) = (1 - B_j)y_j + B_j\mu,$$

   where $B_j = \tau^{-2}/(\tau^{-2} + \sigma_j^{-2})$ is the size of the shrinkage of y_j toward the mean μ. For all observations, compute the shrinkage size $E(B_j|y)$ from the simulated draws of the hyperparameters. Rank the schools from the largest shrinkage to the smallest shrinkage and explain why there are differences.

   b) School A had the largest observed coaching effect of 28. From the simulated draws from the joint distribution of $\theta_1, ..., \theta_J$, compute the posterior probability $P(\theta_1 > \theta_j)$ for $j = 2, ..., J$.

3. **Binomial/beta exchangeable model**

   In Chapter 5, we described the problem of simultaneously estimating the rates of death from stomach cancer for males at risk in the age bracket 45–64 for the largest cities in Missouri. The dataset is available as `cancermortality` in the LearnBayes package. Assume that the numbers of cancer deaths $\{y_j\}$ are independent, where y_j is binomial with sample size n_j and probability of death p_j. To model a prior belief of exchangeability, it is assumed that $p_1, ..., p_{20}$ are a random sample from a beta distribution with parameters a and b. We reparameterize the beta parameters a and b to new values

$$\eta = \frac{a}{a+b}, \quad K = a + b.$$

The hyperparameter η is the prior mean of each p_j and K is a precision parameter. At the last stage of this model, we assign (η, K) the noninformative proper prior

$$g(\eta, K) = \frac{1}{(1+K)^2}, \quad 0 < \eta < 1, K > 0.$$

Due to the conjugate form of the prior, one can derive the following posterior distributions.

- Conditional on the values of the hyperparameters η and K, the probabilities $p_1, ..., p_{20}$ are independent, with p_j distributed beta with parameters $a_j = K\eta + y_j$ and $b_j = K(1 - \eta) + n_j - y_j$.
- The marginal posterior density of (η, K) has the form

$$g(\eta, K|y) \propto \frac{1}{(1+K)^2} \prod_{j=1}^{20} \frac{B(K\eta + y_j, K(1 - \eta) + n_j - y_j)}{B(K\eta, K(1 - \eta))},$$

where $K > 0$ and $0 < \eta < 1$.

a) To summarize the posterior distribution of the hyperparameters η and K, first transform the parameters to the real line by the reexpressions $\theta_1 = \log K$ and $\theta_2 = \log(\eta/(1 - \eta))$. Write an R function to compute values of the log posterior of θ_1 and θ_2.

b) Use a simulation algorithm such as Gibbs sampling (function `gibbs`), random walk Metropolis (function `rwmetrop`), or independence Metropolis (function `indepmetrop`) to obtain a sample of size 1000 from the posterior of (θ_1, θ_2). Summarize the posterior distributions of K and η by 90% interval estimates.

c) Using the simulated sample from the marginal posterior of (θ_1, θ_2), simulate 1000 draws from the joint posterior density of the probabilities $p_1, ..., p_{20}$. Summarize the posterior distribution of each p_j by a 90% interval estimate.

4. **Binomial/beta exchangeable model (continued)**
We assume that the sampling algorithm in Exercise 7.3 has been followed and one has simulated a sample of 1000 values from the marginal posterior of the hyperparameters K and m, and also from the posterior densities of $p_1, ..., p_{20}$.

a) Let y_j^* denote the number of cancer deaths of a future sample of size n_j from the jth city in Missouri. Conditional on the probability p_j distribution of y_j^* is binomial(n_j, p_j). For city 1 (with $n_j = 1083$ patients) and city 15 (with $n_j = 53637$ patients), simulate a sample of 1000 values from the posterior predictive distribution of y_j^*.

b) For cities 1 and 15, the observed numbers of cancer deaths were 0 and 54, respectively. By comparing the observed values of y_j against the respective predictive distributions, decide if these values are consistent with the binomial/beta exchangeable model.

8

Model Comparison

8.1 Introduction

In this chapter we illustrate the use of R to compare models from a Bayesian perspective. We introduce the notion of a Bayes factor in the setting where one is comparing two hypotheses about a parameter. In the setting where one is testing hypotheses about a population mean, we illustrate the computation of Bayes factors in both the one-sided and two-sided settings. We then generalize to the setting where one is comparing two Bayesian models, each consisting of a choice of prior and sampling density. In this case, the Bayes factor is the ratio of the marginal densities for the two models. We illustrate Bayes factor computations in two examples. In the analysis of hitting data for a baseball player, one wishes to compare a "consistent" model with a "streaky" model where the probability of a success may change over a season. In the second application, we illustrate the computation of Bayes factors against independence in a two-way contingency table.

8.2 Comparison of Hypotheses

To introduce Bayesian measures of evidence, suppose one observes Y from a sampling distribution $f(y|\theta)$ and one wishes to test the hypotheses

$$H_0 : \theta \in \Theta_0, \; H_1 : \theta \in \Theta_1,$$

where Θ_0 and Θ_1 form a partition of the parameter space. If one assigns a proper prior density $g(\theta)$, then one can judge the two hypotheses a priori by the prior odds ratio

$$\frac{\pi_0}{\pi_1} = \frac{P(\theta \in \Theta_0)}{P(\theta \in \Theta_1)} = \frac{\int_{\Theta_0} g(\theta)d\theta}{\int_{\Theta_1} g(\theta)d\theta}.$$

After data $Y = y$ is observed, one's beliefs about the parameter are updated by the posterior density

$$g(\theta|y) \propto L(\theta)g(\theta),$$

where $L(\theta)$ is the likelihood function. One's new beliefs about the two hypotheses are summarized by the posterior odds ratio

$$\frac{p_0}{p_1} = \frac{P(\theta \in \Theta_0|y)}{P(\theta \in \Theta_1|y)} = \frac{\int_{\Theta_0} g(\theta|y)d\theta}{\int_{\Theta_1} g(\theta|y)d\theta}.$$

The Bayes factor is the ratio of the posterior odds to the prior odds of the hypotheses

$$BF = \frac{\text{posterior odds}}{\text{prior odds}} = \frac{p_0/p_1}{\pi_0/\pi_1}.$$

The statistic BF is a measure of the evidence provided by the data in support of the hypothesis H_0. The posterior probability of the hypothesis H_0 can be expressed as a function of the Bayes factor and the prior probabilities of the hypotheses by

$$p_0 = \frac{\pi_0 BF}{\pi_0 BF + 1 - \pi_0}.$$

8.3 A One-Sided Test of a Normal Mean

In an example from chapter 14 of Berry (1996), the author was interested in determining his true weight from a variable bathroom scale. We assume the measurements are normally distributed with mean μ and standard deviation σ. The author weighs himself 10 times with the measurements (in pounds) 182, 172, 173, 176, 176, 180, 173, 174, 179, and 175. For simplicity, assume that he knows the accuracy of the scale and $\sigma = 3$ pounds.

If we let μ denote the author's true weight, suppose he is interested in assessing if his true weight is larger than 175 pounds. He wishes to test the hypotheses

$$H_0 : \mu \leq 175, \ H_1 : \mu > 175.$$

Suppose the author has little prior knowledge about his true weight and so he assigns μ a normal prior with mean 170 and standard deviation 5

$$\mu \text{ distributed } N(170, 5).$$

The prior odds of the null hypothesis H_0 is given by

$$\frac{\pi_0}{\pi_1} = \frac{P(\mu \leq 175)}{P(\mu > 175)}.$$

We compute this prior odds from the N(170, 5) density using the `pnorm` function. In the following output, `pmean` and `pvar` are, respectively, the prior mean and prior variance of μ.

```
> pmean=170; pvar=25
> probH=pnorm(175,pmean,sqrt(pvar))
> probA=1-probH
> prior.odds=probH/probA
> prior.odds
```

```
[1] 5.302974
```

So a priori the null hypothesis is five times more likely than the alternative hypothesis.

We enter the 10 weight measurements into R and compute the sample mean $\bar{y}$ and the associated sampling variance `sigma2` equal to σ^2/n.

```
> weights=c(182, 172, 173, 176, 176, 180, 173, 174, 179, 175)
> ybar=mean(weights)
> sigma2=3^2/length(weights)
```

By the familiar normal density/normal prior updating formula, the posterior precision (inverse of the variance) of μ is the sum of the precisions of the data and the prior.

```
> post.precision=1/sigma2+1/pvar
> post.var=1/post.precision
```

The posterior mean of μ is the weighted sum of the sample mean and the prior mean where the weights are proportional to the respective precisions.

```
> post.mean=(ybar/sigma2+pmean/pvar)/post.precision
> c(post.mean,sqrt(post.var))
```

```
[1] 175.7915058   0.9320547
```

The posterior density of μ is N(175.79, 0.93).

Using this normal posterior density, we calculate the odds of the null hypothesis.

```
> post.odds=pnorm(175,post.mean,sqrt(post.var))/
+  (1-pnorm(175,post.mean,sqrt(post.var)))
> post.odds
```

```
[1] 0.2467017
```

So the Bayes factor in support of the null hypothesis is

```
> BF = post.odds/prior.odds
> BF
```

```
[1] 0.04652139
```

From the prior probabilities and the Bayes factor, we can compute the posterior probability of the null hypothesis.

values (sample mean, sample size, known sampling standard deviation). Since
it may be difficult to assess values for τ, the function allows the user to input
a vector of plausible values.

The R code for the computation in this example is shown here. Note the
values .5, 1, 2, 4, and 8 are inputted as possible values for τ.

```
> weights=c(182, 172, 173, 176, 176, 180, 173, 174, 179, 175)
> data=c(mean(weights),length(weights),3)
> t=c(.5,1,2,4,8)
> mnormt.twosided(170,.5,t,data)
$bf
[1] 1.462146e-02 3.897038e-05 1.894326e-07 2.591162e-08
[5] 2.309739e-08
$post
[1] 1.441076e-02 3.896887e-05 1.894325e-07 2.591162e-08
[5] 2.309739e-08
```

For each value of the prior standard deviation τ, the program gives the
Bayes factor in support of the hypothesis that μ takes on the specific value
and the posterior probability that the hypothesis H is true. If the author uses
a normal (170, 2) density to reflect alternative values for his weight μ, then
the Bayes factor in support of the hypothesis $\mu = 170$ is equal to .0000002.
The posterior probability that his weight hasn't changed is .0000002, which
is much smaller than the author's prior probability of .5. He should conclude
that his current weight is not 170.

8.5 Comparing Two Models

The Bayesian approach to comparing hypotheses can be generalized to compare two models. If we let y denote the vector of data and θ the parameter,
then a Bayesian model consists of a specification of the sampling density $f(y|\theta)$
and the prior density $g(\theta)$. Given this model, one can compute the marginal
or prior predictive density of the data

$$m(y) = \int f(y|\theta)g(\theta)d\theta.$$

Suppose we wish to compare two Bayesian models

$$M_0 : y \sim f_1(y|\theta_0), \theta_0 \sim g_1(\theta_0), \quad M_1 : y \sim f_2(y|\theta_1), \theta_1 \sim g_2(\theta_1),$$

where it is possible that the definition of the parameter θ may differ between
models. Then the Bayes factor in support of model M_0 is the ratio of the
respective marginal densities (or prior predictive densities) of the data for the
two models.

$$BF = \frac{m_0(y)}{m_1(y)}.$$

If π_0 and π_1 denote the respective prior probabilities of the models M_0 and M_1, then the posterior probability of model M_0 is given by

$$P(M_0|y) = \frac{\pi_0 BF}{\pi_0 BF + \pi_1}.$$

A simple way of approximating a marginal density is by Laplace's method, described in Section 3 of Chapter 5. Let $\hat{\theta}$ denote the posterior mode and $H(\theta)$ denote the Hessian (second derivative matrix) of the log posterior density. Then the prior predictive density can be approximated as

$$m(y) \approx (2\pi)^{d/2} g(\hat{\theta}) f(y|\hat{\theta})| - H(\hat{\theta})|^{1/2},$$

where d is the number of parameters. On the log scale, we have

$$\log m(y) \approx (d/2) \log(2\pi) + \log(g(\hat{\theta}) f(y|\hat{\theta})) + (1/2) \log | - H(\hat{\theta})|.$$

Once an R function is written to compute the logarithm of the product $f(y|\theta)g(\theta)$, then the function `laplace` can be applied and the component of the output `int` gives an estimate of $\log m(y)$. By applying this method for several models, one can use the computed values of $m(y)$ to compute a Bayes factor.

8.6 Models for Soccer Goals

To illustrate the use of the function `laplace` in computing Bayes factors, suppose you are interested in learning about the mean number of goals scored by a team in Major League Soccer. You observe the number of goals scored $y_1, ..., y_n$ for n games. Since goals are relatively rare events, it is reasonable to assume that the y_is are distributed according to a Poisson distribution with mean λ. We consider the use of the following four subjective priors for λ:

1. **Prior 1.** You assign a conjugate gamma prior to λ of the form

$$g(\lambda) \propto \lambda^{\alpha-1} \exp\{-\beta\lambda\}, \lambda > 0,$$

with $\alpha = 4.57$ and $\beta = 1.43$. This prior says that you believe that a team averages about 3 goals a game and the quartiles for λ are given by 2.10 and 4.04.

2. **Prior 2.** It is more convenient for you to represent prior opinion in terms of symmetric distributions. So you assume that $\log \lambda$ is normal with mean 1 and standard deviation .5. The quartiles of this prior for $\log \lambda$ are 0.66 and 1.34, which translates to prior quartiles for λ of 1.94 and 3.81. Note that Prior 1 and this prior reflect similar beliefs about the location of the mean rate λ.

3. **Prior 3.** This prior assumes that $\log \lambda$ is $N(2, .5)$. The prior quartiles for the rate λ are 5.27 and 10.35. This prior says that you believe teams score a lot of goals in Major League Soccer.
4. **Prior 4.** This prior assumes that $\log \lambda$ is $N(1, 2)$ with associated quartiles for the rate λ of 1.92 and 28.5. This prior reflects little knowledge about the scoring pattern of soccer games.

The number of goals were observed for a particular team in Major League Soccer for the 2006 season. The dataset is available as `soccergoals` in the LearnBayes package. The likelihood of λ, assuming the Poisson model, is given by

$$L(\lambda) = \frac{\exp(-n\lambda)\lambda^s}{\prod_{i=1}^{n} y_i!},$$

where $s = \sum_{i=1}^{n} y_i$. For our dataset, $n = 35$ and $s = 57$. Fig. 8.1 displays the likelihood on the $\log \lambda$ scale together with the four proposed priors described earlier. Priors 1 and 2 seem pretty similar in location and shape. We see substantial conflict between the likelihood and Prior 3, and the shape of Prior 4 is very flat relative to the likelihood.

To use the function `laplace`, we have to write short functions defining the log posterior. The first function `logpoissgamma` computes the log posterior with Poisson sampling and a gamma prior. Following our usual strategy, we transform λ to the real-valued parameter $\theta = \log \lambda$. The arguments to the function are `theta` and `datapar`, a list that contains the data vector `data` and the parameters of the gamma prior `par`. Note that we use the R function `dgamma` in computing both the likelihood and the prior.

```
logpoissgamma=function(theta,datapar)
{
y=datapar$data
npar=datapar$par
lambda=exp(theta)
loglike=log(dgamma(lambda,shape=sum(y)+1,rate=length(y)))
logprior=log(dgamma(lambda,shape=npar[1],rate=npar[2])*lambda)
return(loglike+logprior)
}
```

Similarly, we write the function `logpoissnormal` to compute the log posterior of $\log \lambda$ for Poisson sampling and a normal prior. This function uses both the R functions `dgamma` and `dnorm`.

```
logpoissnormal=function(theta,datapar)
{
y=datapar$data
npar=datapar$par
```

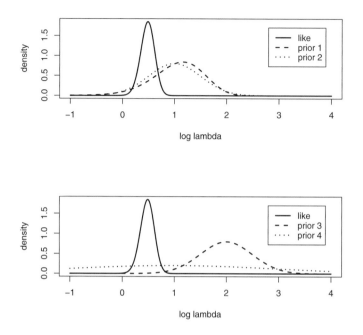

Fig. 8.1. The likelihood function and four priors on $\theta = \log \lambda$ for the soccer goal example.

```
lambda=exp(theta)
loglike=log(dgamma(lambda,shape=sum(y)+1,scale=1/length(y)))
logprior=log(dnorm(theta,mean=npar[1],sd=npar[2]))
return(loglike+logprior)
}
```

We first load in the datafile `soccergoals`; there is one variable `goals` in this dataset that we make available by the `attach` command. For each of the four priors, we use the function `laplace` to summarize the posterior. If the output of the function is `fit`, `fit$mode` is the posterior mode, `fit$var` is the associated estimate at the posterior variance and `fit$int` is the estimate at $\log m(y)$.

```
> data(soccergoals)
> attach(soccergoals)
> datapar=list(data=goals,par=c(4.57,1.43))
> fit1=laplace(logpoissgamma,.5,10,datapar)
> datapar=list(data=goals,par=c(1,.5))
> fit2=laplace(logpoissnormal,.5,10,datapar)
```

```
> datapar=list(data=goals,par=c(2,.5))
> fit3=laplace(logpoissnormal,.5,10,datapar)
> datapar=list(data=goals,par=c(1,2))
> fit4=laplace(logpoissnormal,.5,10,datapar)
```

We display the posterior modes, posterior standard deviations, and log marginal densities for the four models corresponding to the four priors.

```
> postmode=c(fit1$mode,fit2$mode,fit3$mode,fit4$mode)
> postsd=sqrt(c(fit1$var,fit2$var,fit3$var,fit4$var))
> logmarg=c(fit1$int,fit2$int,fit3$int,fit4$int)
> cbind(postmode,postsd,logmarg)

        postmode      postsd    logmarg
[1,]  0.5247821  0.1274428  -1.502965
[2,]  0.5207796  0.1260714  -1.255170
[3,]  0.5825327  0.1224715  -5.076322
[4,]  0.4899378  0.1320168  -2.137214
```

By use of the values of $\log m(y)$, one can compare the different models by Bayes factors. Does it matter if we use a gamma(4.57, .7) prior on λ or a normal(1, .5) prior on $\log \lambda$? To answer this question, we can compute the Bayes factor in support of Prior 2 over Prior 1:

$$BF_{21} = \frac{m_2(y)}{m_1(y)} = \exp(-1.255170 + 1.502965) = 1.28.$$

There is slight support for the normal prior – this makes sense from Fig. 8.1 since Prior 2 is slightly closer to the likelihood function. Comparing Prior 2 with Prior 3, the Bayes factor in support of Prior 2 is

$$BF_{23} = \frac{m_2(y)}{m_3(y)} = \exp(-1.255170 + 5.076322) = 45.6,$$

indicating large support for Prior 2. Actually, note that the locations of the likelihood and Prior 3 are far apart, indicating a conflict between the data and the prior and a small value of $m_3(y)$. Comparing Prior 2 with Prior 4, the Bayes factor in support of Prior 2 is

$$BF_{24} = \frac{m_2(y)}{m_4(y)} = \exp(-1.255170 + 2.137214) = 2.42.$$

Generally, the marginal probability for a prior will decrease as the prior density becomes more diffuse.

8.7 Is a Baseball Hitter Really Streaky?

In sports, we observe much streaky behavior in players and teams. For example, in the sport of baseball, one measure of success of a hitter is the batting average or proportion of base hits. During a baseball season, there will be periods when a player is "hot" and has an unusually high batting average, and there will also be periods when the player is "cold" and has a very small batting average. We observe many streaky patterns in the performance of players. The interesting question is what this streaky data says about the *ability* of a player to be streaky.

In baseball, the player has opportunities to bat in an individual season – we call these opportunities "at-bats." In each at-bat, there are two possible outcomes – a hit (a success) or an out (a failure). Suppose we divide all of the at-bats in a particular baseball season into N periods. Let p_i denote the probability the player gets a hit in a single at-bat during the ith period, $i = 1, ..., N$. If a player is truly consistent or nonstreaky, then the probability of a hit stays constant across all periods; we call this the nonstreaky model M_0:

$$M_0 : p_1 = ... = p_N = p.$$

To complete this model specification, we assign the common probability value p a uniform prior on $(0, 1)$.

On the other hand, if the player is truly streaky, then the probability of a hit p_i will change across the season. A convenient way to model this variation in the probabilities is to assume that $p_1, ..., p_N$ are a random sample from a beta density of the form

$$g(p) = \frac{1}{B(K\eta, K(1 - \eta))} p^{K\eta - 1} (1 - p)^{K(1 - \eta) - 1}, 0 < p < 1.$$

In the density g, η is the mean and K is a precision parameter. We can index this streaky model by the parameter K; we represent the streaky model by

$$M_K : p_1, ..., p_N \text{ iid beta}(K\eta, K(1 - \eta)).$$

For this model, we place a uniform prior on the mean parameter η, reflecting little knowledge about the location of the random effects distribution. Note that as the precision parameter K approaches infinity, the streaky model M_K approaches the consistent model M_0.

To compare the models M_0 and M_K, we need to compute the associated marginal densities. Under the model M_0, the numbers of hits $y_1, ..., y_N$ are independent, where y_i is binomial(n_i, p). With the assumption that p is uniform$(0, 1)$, we obtain the marginal density

$$m_0(y) = \int \prod_{i=1}^{N} \binom{n_i}{y_i} p^{y_i} (1 - p)^{n_i - y_i} dp$$

$$= \prod_{i=1}^{N} \binom{n_i}{y_i} B(\sum_{i=1}^{N} y_i + 1, \sum_{i=1}^{N} (n_i - y_i) + 1).$$

Under the alternative "streaky" model, the marginal density is given by

$$m_K(y) = \int \prod_{i=1}^{N} \binom{n_i}{y_i} p_i^{y_i}(1-p_i)^{n_i-y_i} \frac{p_i^{K\eta-1}(1-p_i)^{K(1-\eta)-1}}{B(K\eta,K(1-\eta))} dp_1...dp_N$$

$$= \prod_{i=1}^{N} \binom{n_i}{y_i} \int_0^1 \frac{\prod_{i=1}^{N} B(y_i+K\eta,n_i-y_i+K(1-\eta))}{B(K\eta,K(1-\eta))^N} d\eta.$$

The Bayes factor in support of the "streaky" model H_K compared to the "nonstreaky" model H_0 is given by

$$B_K = \frac{m_K(y)}{m_0(y)}$$

$$= \frac{1}{B(\sum y_i+1,\sum(n_i-y_i)+1)} \int_0^1 \frac{\prod_{i=1}^{N} B(y_i+K\eta,n_i-y_i+K(1-\eta))}{B(K\eta,K(1-\eta))^N} d\eta.$$

We use the function `laplace` to compute the integral in the Bayes factor B_K. We first transform the variable η in the integral to the real-valued variable $\theta = \log(\eta/(1-\eta))$. Using the R function `lbeta` that computes the logarithm of the beta function, we define the following function `bfexch` that computes the log integral. The inputs to this function are `theta` and a list `datapar` with components `data` (a matrix with columns y and n) and K.

```
bfexch=function(theta,datapar)
{
y=datapar$data[,1]; n=datapar$data[,2]; K=datapar$K
eta=exp(theta)/(1+exp(theta))
N=length(y)
z=0*theta;
for (i in 1:N)
   z=z+lbeta(K*eta+y[i],K*(1-eta)+n[i]-y[i])
z=z-N*lbeta(K*eta,K*(1-eta))+log(eta*(1-eta))
z=z-lbeta(sum(y)+1,sum(n-y)+1)
return(z)
}
```

To compute the Bayes factor B_K for a specific value, say K0, we use the function `laplace` with inputs the function `bfexch`, a starting value of $\eta = 0$, 10 iterations of Newton's method, and the list `datapar` using the value K0.

```
s=laplace(bfexch,0,10,list(data=data,K=K0)))
```

The list s is the output of `laplace`; the component `s$int` gives the estimate at the logarithm of the Bayes factor $\log B_K$.

To illustrate the use of this method, we consider the hitting data for the New York Yankee player Derek Jeter for the 2004 baseball season. Jeter was one of the "star" players on this team, and he experienced an unusual hitting slump during the early part of the season that attracted much attention from the local media.

Hitting data for Jeter were collected for each of the 155 games he played in that particular season. A natural way of defining periods during the season is by games, so $N = 155$. However, it is difficult to detect streakiness in these hitting data since Jeter only had about 4–5 opportunities to hit in each game. So we group the data into five-game intervals. The original game-by-game data are available as `jeter2004` in the LearnBayes package. In the following R code, we read in the complete hitting data for Jeter and use the `regroup` function to group the data into periods of five games.

```
> data(jeter2004)
> attach(jeter2004)
> data=cbind(H,AB)
> data1=regroup(data,5)
```

The matrix `data1` contains the grouped hitting data $(y_i, n_i), i = 1, ..., 31$, where y_i is the number of hits by Jeter in n_i at-bats in the ith interval of games. These data are listed in Table 8.1.

Table 8.1. Hitting data from Derek Jeter for 2004 baseball season.

Period	(y, n)	Period	(y, n)	Period	(y, n)	Period	(y, n)
1	(4, 19)	9	(6, 24)	17	(5, 21)	25	(5, 20)
2	(6, 22)	10	(12,24)	18	(6, 21)	26	(4, 17)
3	(4, 22)	11	(4, 15)	19	(4, 23)	27	(11, 20)
4	(0, 20)	12	(11,21)	20	(8, 19)	28	(7, 21)
5	(5, 22)	13	(5 ,21)	21	(8, 21)	29	(9, 21)
6	(3, 19)	14	(8, 21)	22	(6, 23)	30	(6, 20)
7	(8, 24)	15	(7, 18)	23	(3, 22)	31	(7, 19)
8	(3, 23)	16	(7, 22)	24	(6, 18)		

We compute the Bayes factor for a sequence of values of $\log K$ using the function `laplace` and the definition of the log integral defined in the function `bfexch`. In this example, the vector `logK` contains the values $\log(K) = 2, 3, 4, 5$, and 6 and the vector `logBF` stores the corresponding values of the log Bayes factor $\log B_K$. We display in a matrix the values of $\log K$, the values of K, the values of $\log B_K$, and the values of the Bayes factor B_K.

```
> logK=seq(2,6)
> logBF=0*logK
> for (j in 1:length(logK))
+ {
+ s=laplace(bfexch,0,10,list(data=data1,K=exp(logK[j])))
+ logBF[j]=s$int
+ }
> cbind(logK,exp(logK),logBF,exp(logBF))
```

```
      [,1]        [,2]          [,3]        [,4]
[1,]    2    7.389056  -2.9441182  0.05264847
[2,]    3   20.085537   1.0482048  2.85252569
[3,]    4   54.598150   1.4380139  4.21232144
[4,]    5  148.413159   0.8160944  2.26164940
[5,]    6  403.428793   0.3538964  1.42460759
```

We see from the output that the value $\log K = 4$ is most supported by the data with a corresponding Bayes factor of $B_K = 4.21$. This particular streaky model is approximately four times as likely as the consistent model. This indicates that Jeter indeed did display some true streakiness in his hitting behavior for this particular baseball season.

8.8 A Test of Independence in a Two-Way Contingency Table

A basic problem in statistics is to explore the relationship between two categorical measurements. To illustrate this situation, consider the following example presented in Moore (1995) in which North Carolina State University looked at student performance in a course taken by chemical engineering majors. Researchers wished to learn about the relationship between the time spent in extracurricular activities and the grade in the course. Data on these two categorical variables were collected from 119 students, and the responses are presented using the contingency table in Table 8.2.

Table 8.2. Two-way table relating student performance and time spent in extracurricular activities.

	Extracurricular Activities (hr per week)		
	< 2	2 to 12	> 12
C or better	11	68	3
D or F	9	23	5

To learn about the possible relationship between participation in extracurricular activities and grade, one tests the hypothesis of independence. The usual non-Bayesian approach of testing the independence hypothesis is based on a Pearson chi-squared statistic that contrasts the observed counts with expected counts under an independence model. In R, we read in the table of counts and use the function chisq.test to test the independence hypothesis:

```
> data=matrix(c(11,9,68,23,3,5),c(2,3))
> data
```

```
       [,1] [,2] [,3]
[1,]    11   68    3
[2,]     9   23    5

> chisq.test(data)

        Pearson's Chi-squared test

data:   data
X-squared = 6.9264, df = 2, p-value = 0.03133

Warning message:
Chi-squared approximation may be incorrect in: chisq.test(data)
```

Here the p-value is approximately .03, which is some evidence that one's grade is related to the time spent on extracurricular activities.

From a Bayesian viewpoint, there are two possible models – the model M_I that the two categorical variables are independent and the model M_D that the two variables are dependent in some manner. To describe the Bayesian models, assume that these data represent a random sample from the population of interest and the counts of the table have a multinomial distribution with proportion values as shown in Table 8.3. Under the dependence model M_D, the proportion values $p_{11}, \ldots, p_{23}$ can take any values that sum to 1, and we assume the prior density places a uniform distribution over this space.

Table 8.3. Probabilities of the table under the hypothesis of dependence.

	Extracurricular Activities (hr per week)		
	< 2	2 to 12	> 12
C or better	p_{11}	p_{12}	p_{13}
D or F	p_{21}	p_{22}	p_{23}

Under the independence model M_I, the proportions in the table are determined by the marginal probabilities $\{p_{1+}, p_{2+}\}$ and $\{p_{+1}, p_{+2}, p_{+3}\}$ as displayed in Table 8.4. Here the unknown parameters are the proportions of students in different activity levels and the proportions with different grades. We assume that our knowledge about these two sets of proportions, $\{p_{i+}\}$ and $\{p_{+j}\}$, are independent and assign to each set a uniform density over all possible values.

We have defined two models – a dependence model M_D where the multinomial proportions are uniformly distributed and an independence model M_I where the multinomial proportions have an independence structure and the marginal proportions are assigned independent uniform priors. It can be shown

Table 8.4. Probabilities of the table under the hypothesis of independence.

	Extracurricular Activities (hr per week)			
	< 2	2 to 12	> 12	
C or better	$p_1 + p_{+1}$	$p_1 + p_{+2}$	$p_1 + p_{+3}$	p_{1+}
D or F	$p_2 + p_{+1}$	$p_2 + p_{+2}$	$p_2 + p_{+3}$	p_{2+}
	p_{+1}	p_{+2}	p_{+3}	

that the Bayes factor in support of the dependence model over the independence model is given by

$$BF = \frac{D(y+1)D(1_R)D(1_C)}{D(1_{RC})D(y_R+1)D(y_C+1)},$$

where y is the matrix of counts, y_R is the vector of row totals, y_C is the vector of column totals, 1_R is the vector of ones of length R, and $D(\nu)$ is the Dirichlet function defined by

$$D(\nu) = \prod \Gamma(\nu_i)/\Gamma(\sum \nu_i).$$

The R function `ctable` will compute this Bayes factor for a two-way contingency table. One inputs a matrix a of prior parameters for the matrix of probabilities. By taking a matrix a of ones, one is assigning a uniform prior on $\{p_{ij}\}$ and uniform priors on $\{p_{i+}\}$ and $\{p_{+j}\}$ under the dependence model. The output of this problem is the value of the Bayes factor. Here the value is $BF = 1.66$, which indicates modest support against independence.

```
> a=matrix(rep(1,6),c(2,3))
> a

     [,1] [,2] [,3]
[1,]    1    1    1
[2,]    1    1    1

> ctable(data,a)

[1] 1.662173
```

We are comparing "uniform" with "independence" models for a contingency table. One criticism of this method is that we may not really be interested in a "uniform" alternative model. Perhaps we would like to compare "independence" with a model where the cell probabilities are "close to independence." Such a model was proposed by Albert and Gupta (1981). Suppose the table probabilities $\{p_{ij}\}$ are assigned a conjugate Dirichlet distribution of the form

$$g(p) \propto \prod p_{ij}^{K\eta_{ij}-1},$$

where the prior means $\{\eta_{ij}\}$ satisfy an independence configuration

$$\eta_{ij} = \eta_i^A \eta_j^B.$$

This structure of prior means is illustrated for our example in Table 8.5. Then the vectors of prior means of the margins $\{\eta_i^A\}$, and $\{\eta_j^B\}$ are assigned uniform distributions. This model will be labeled M_K as it is indexed by the Dirichlet precision parameter K. As K approaches infinity, the model approaches the independence hypothesis M_I, where the marginal probabilities have uniform distributions.

Table 8.5. Prior means of the cell probabilities of the table under "close to independence" model.

	Extracurricular Activities (hr per week)			
	< 2	2 to 12	> 12	
C or better	$\eta_1^A \eta_1^B$	$\eta_1^A \eta_2^B$	$\eta_1^A \eta_3^B$	η_1^A
D or F	$\eta_2^A \eta_1^B$	$\eta_2^A \eta_2^B$	$\eta_2^A \eta_3^B$	η_2^A
	η_1^B	η_2^B	η_3^B	

It can be shown that the Bayes factor in support of the "close to independence" model M_K over the independence model M_I is given by

$$BF_K = \frac{1}{D(y_R + 1)D(y_C + 1)} \int \frac{D(K\eta^A \eta^B + y)}{D(K\eta^A \eta^B)} d\eta^A d\eta^B,$$

where $K\eta^A \eta^B + y$ is the vector of values $\{K\eta_i^A \eta_j^B + y_{ij}\}$ and the integral is taken over the vectors of marginal prior means $\eta^A = \{\eta_i^A\}$ and $\eta^B = \{\eta_j^B\}$.

One straightforward way of computing the Bayes factor is by importance sampling. The Bayes factor can be represented as the integral

$$BF_K = \int h(\theta) d\theta,$$

where $\theta = (\eta^A, \eta^B)$. Suppose the integrand can be approximated by the density $g(\theta)$, where g is easy to simulate. Then by writing the integral as

$$BF_K = \int \frac{h(\theta)}{g(\theta)} g(\theta) d\theta,$$

we can approximate the integral as

$$BF_K \approx \frac{\sum_{j=1}^m h(\theta_j)/g(\theta_j)}{m},$$

where $\theta_1, ..., \theta_m$ are independent simulated draws from $g(\theta)$. The simulation standard error of this importance sampler estimate is given by

$$se = \text{standard deviation}\left(\{h(\theta_j)/g(\theta_j)\}\right)/\sqrt{m}.$$

In our example, it can be shown, as K approaches infinity, the posterior of the vectors of marginal prior means η^A and η^B can be shown to be independent with

$$\eta^A \text{distributed } Dirichlet(y_R + 1), \quad \eta^B \text{distributed } Dirichlet(y_C + 1),$$

where the Dirichlet distribution on the vector η with parameter vector a has the density proportional to $\prod \eta_i^{a_i-1}$. This density is a convenient choice for importance sampler since it is easy to simulate draws from a Dirichlet distribution.

Using this importance sampling algorithm, the function bfindep computes the Bayes factor using this alternative "close to independence" model. One inputs the data matrix y, the Dirichlet precision parameter K, and the size of the simulated sample m. The output is a list with two components: bf, the value of the Bayes factor, and nse, an estimate at the simulation standard error of the computed value of BF.

In the following R input, we compute the Bayes factor for a sequence of values of log K. The output gives the value of the log Bayes factor and the Bayes factor for some values of log K. Fig. 8.2 displays the log Bayes factor as a function of log K and 10,000 simulation draws. (We used the R function spm in the SemiPar package to smooth out the simulation errors in the computed log Bayes factors before plotting.) Note that this maximum value of the Bayes factor is 2.3, indicating some support for an alternative model that is in the neighborhood of the independence model.

```
> logK=seq(2,7,by=.2)
> logBF=0*logK
> for (j in 1:length(logK))
+ {x=bfindep(data,exp(logK[j]),100000); logBF[j]=log(x$bf)}
> cbind(logK,logBF,exp(logBF))
```

```
        logK        logBF
 [1,]    2.0  -1.53308341 0.2158690
 [6,]    3.0   0.04157343 1.0424497
[11,]    4.0   0.83315205 2.3005588
[16,]    5.0   0.73923892 2.0943409
[21,]    6.0   0.43584433 1.5462681
[26,]    7.0   0.19970982 1.2210484
```

8.9 Further Reading

Carlin and Louis (2000), chapter 6, and Kass and Raftery (1995) provide general discussions of the use of Bayes factors in selecting models. Berger

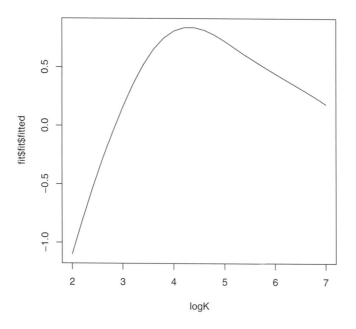

Fig. 8.2. Plot of log Bayes factor in support of model M_K over M_I against the precision parameter $\log K$.

and Sellke (1987) and Casella and Berger (1987) describe the relationship between Bayesian and frequentist measures of evidence in the two-sided and one-sided testing situations, respectively. Gunel and Dickey (1974) describe the use of Dirichlet distributions in the development of tests for contingency tables, and Albert and Gupta (1981) introduce the use of mixtures of Dirichlet distributions for contingency tables.

8.10 Summary of R Functions

bfexch – computes the logarithm of the integral of the Bayes factor for testing homogeneity of a set of probabilities
Usage: bfexch(theta,datapar)
Arguments: theta, vector of values of the logit of the prior hyperparameter η; datapar, list with components data (matrix with columns y and n) and K (prior precision hyperparameter)
Value: vector of values of the logarithm of the integral

`bfindep` – computes a Bayes factor against independence for a two-way contingency table assuming a "close to independence" alternative model
Usage: `bfindep(y, K, m)`
Arguments: `y`, matrix of counts; `K`, Dirichlet precision hyperparameter; `m`, number of simulations
Value: `bf`, value of the Bayes factor against independence; `nse`, estimate of the simulation standard error of the computed value of the Bayes factor

`ctable` – computes a Bayes factor against independence for a two-way contingency table assuming uniform prior distributions
Usage: `ctable(y,a)`
Arguments: `y`, matrix of counts; `a`, matrix of prior parameters for the matrix of probabilities
Value: the Bayes factor against the hypothesis of independence

`logpoissgamma` – computes the logarithm of the posterior with Poisson sampling and a gamma prior
Usage: `logpoissgamma(theta, datapar)`
Arguments: `theta`, vector of values of the log mean parameter; `datapar`, list with components `data` (vector of sample values) and `par` (vector of parameters of the gamma prior)
Value: value of the log posterior for all values in theta

`logpoissnormal` – computes the logarithm of the posterior with Poisson sampling and a normal prior
Usage: `logpoissnormal(theta, datapar)`
Arguments: `theta`, vector of values of the log mean parameter; `datapar`, list with components `data` (vector of sample values) and `par` (vector of parameters of the normal prior)
Value: value of the log posterior for all values in theta

`mnormt.onesided` – Bayesian test of the hypothesis that a normal mean M is less than or equal to a specific value
Usage: `mnormt.onesided(mu0,normpar,data)`
Arguments: `mu0`, value of the normal mean to be tested; `normpar`, vector of mean and standard deviation of the normal prior distribution; `data`, vector of sample mean, sample size, and known value of the population standard deviation
Value: `BF`, Bayes factor in support of the null hypothesis; `prior.odds`, the prior odds of the null hypothesis; `post.odds`, the posterior odds of the null hypothesis, `postH`, the posterior probability of the null hypothesis

`mnormt.twosided` – Bayesian test of the hypothesis that a normal mean M is equal to a specific value
Usage: `mnormt.twosided(mu0, probH, tau, data)`
Arguments: `mu0`, the value of the normal mean to be tested; `probH`, the prior probability of the null hypothesis; `tau`, vector of values of the prior standard

deviation under the alternative hypothesis; data, vector of sample mean, sample size, and known value of the population standard deviation

Value: bf, vector of values of the Bayes factor in support of the null hypothesis; post, vector of values of the posterior probability of the null hypothesis

8.11 Exercises

1. **A one-sided test of a binomial probability**

   In 1986, the *St. Louis Post Dispatch* was interested in measuring public support for the construction of a new indoor stadium. The newspaper conducted a survey in which they interviewed 301 registered voters. Let p denote the proportion of all registered voters in the St. Louis voting district opposed to the stadium. A city councilman wishes to test the hypotheses $H : p \geq .5$, $K : p < .5$.

   a) The number y opposed to the stadium construction is assumed binomial$(301, p)$. Suppose the survey result is $y = 135$. Using the R function pbinom, compute the p-value $P(y \leq 135|p = .5)$. If this probability is small, say under 5%, then one concludes that there is significant evidence in support of the hypothesis $K : p < .5$.

   b) Suppose one places a uniform prior on p. Compute the prior odds of the hypothesis K.

   c) After observing $y = 135$, the posterior distribution of p is beta$(136, 167)$. By use of the R function pbeta, compute the posterior odds of the hypothesis K.

   d) Compute the Bayes factor in support of the hypothesis K.

2. **A two-sided test of a normal mean (example from Weiss (2001))**

   For last year, a sample of 50 cell phone users had a mean local monthly bill of \$41.40. Do these data provide sufficient evidence to conclude that last year's mean local monthly bill for cell phone users has changed from the 1996 mean of \$47.70? (Assume that the population standard deviation is $\sigma = \$25$.)

   a) The usual statistic for testing the value of a normal mean μ is $z = \sqrt{n}(\bar{y} - \mu)/\sigma$. Use this statistic and the R function pnorm to compute a p-value for testing the hypothesis $H : \mu = 47.7$.

   b) Suppose one assigns a prior probability of .5 to the null hypothesis. Use the R function mnormt.twosided to compute the posterior probability of H. The arguments to mnormt.twosided are the value to be tested (47.70), the prior probability of H (.5), the standard deviation τ of the prior under the alternative hypothesis (assume $\tau = 4$), and the data vector (values of sample mean, sample size, and known sampling standard deviation).

   c) Compute the posterior probability of H for the alternative values τ =1, 4, 6, 8, and 10. Compare the values of the posterior probability with the value of the p-value computed in part (a).

3. **Comparing Bayesian models by a Bayes factor**

Suppose that the number of births to women during a month at a particular hospital has a Poisson distribution with parameter R. During a given year at a particular hospital, 66 births were recorded in January and 48 births were recorded in April. If the birth rates during January and April are given by R_J and R_A respectively, then (assuming independence) the probability of the sample result is

$$f(data|R_J, R_A) = \frac{e^{-R_J} R_J^{66}}{66!} \frac{e^{-R_A} R_A^{48}}{48!}.$$

Consider the following two priors for (R_J, R_A):

- $M_1 : R_J \sim$ gamma(240, 4), $R_A \sim$ gamma(200, 4).
- M_2: $R_J = R_A$ and the common value of the rate $R \sim$ gamma(220, 4).

a) Write R functions to compute the logarithm of the posterior density of (R_J, R_A) under model M_1, and the logarithm of the posterior density of R under model M_2.

b) Use the function laplace to compute the logarithm of the predictive density for both models M_1 and M_2.

c) Compute the Bayes factor in support of the model M_1.

4. **Is a basketball player streaky?**

Kobe Bryant is one of the most famous players in professional basketball. Shooting data were obtained for Bryant for the first 15 games in the 2006 season. For game i, one records the number of field goal attempts n_i and the number of successful field goals y_i; the data are displayed in Table 8.6. If p_i denotes the probability that Kobe makes a shot during the ith game, it is of interest to compare the nonstreaky hypothesis

$$M_0 : p_1 = \ldots = p_{15} = p, \ p \sim \text{uniform}(0, 1)$$

against the streaky hypothesis that the p_i vary according to a beta distribuiton

$$M_K : p_1, \ldots, p_{15} \text{ random sample from beta}(K\eta, K(1-\eta)), \eta \sim \text{uniform}(0, 1).$$

Use the function laplace together with the function bfexch to compute the logarithm of the Bayes factor in support of the streaky hypothesis M_K. Compute the log of the Bayes factors for values of $K = 10, 20, 50,$ and 100. Based on your work, is there much evidence that Bryant displayed true streakiness in his shooting performance in these 15 games?

5. **Test of independence (example from Agresti and Franklin (2005))**

The 2002 General Social Survey asked the question "Taken all together, would you say that you are very happy, pretty happy, or not too happy?" Also the survey asked "Compared with American families in general, would you say that your family income is below average, average, or above average?" Table 8.7 cross-tabulates the answers to these two questions.

Table 8.6. Shooting data for Kobe Bryant for the first 15 games during the 2006 basketball season.

Game	(y, n)	Game	(y, n)
1	(8, 15)	9	(12, 23)
2	(4, 10)	10	(9, 18)
3	(5, 7)	11	(8, 24)
4	(12, 19)	12	(7, 23)
5	(5, 11)	13	(19, 26)
6	(7, 17)	14	(11, 23)
7	(10, 19)	15	(7, 16)
8	(5, 14)		

Table 8.7. Happiness and family income from 2002 General Social Survey.

	Happiness		
Income	Not Too Happy	Pretty Happy	Very Happy
Above Average	17	90	51
Average	45	265	143
Below Average	31	139	71

a) By use of the Pearson chi-square statistic, use the function `chisq.test` to test the hypothesis that happiness and family income are independent. Based on the p-value, is there evidence to suggest that the level of happiness is dependent on the family income?

b) Consider two models, a "dependence model" where the underlying multinomial probability vector is uniformly distributed and an "independence model" where the cell probabilities satisfy an independence configuration and the marginal probability vectors have uniform distributions. Using the R function `ctable`, compute the Bayes factor in support of the dependence hypothesis.

c) Instead of the analysis in part (b), suppose that one wishes to compare the independence model with the "close to independence" model M_K described in Section 8.8. Using the function `bfindep`, compute the Bayes factor in support of the model M_K for values of $\log K = 2, 3, 4, 5, 6,$ and 7.

d) Compare the frequentist measure of evidence against independence with the Bayesian measures of evidence computed in parts (b) and (c). Which type of measure, frequentist or Bayesian, indicates more evidence against independence?

9

Regression Models

9.1 Introduction

In this chapter, we illustrate R to fit some common regression models from a Bayesian perspective. We first outline the Bayesian normal regression model and describe algorithms to simulate from the joint distribution of regression parameters and error variance and the predictive distribution of future observations. One can judge the adequacy of the fitted model through use of the posterior predictive distribution and the inspection of the posterior distributions of Bayesian residuals. We then illustrate the R Bayesian computations in an example where one is interested in explaining the variation of extinction times of birds in terms of their nesting behavior, their size, and their migrant status. We conclude by illustrating the Bayesian fitting of a survival regression model.

9.2 Normal Linear Regression

9.2.1 The Model

In the usual multiple regression problem, we are interested in describing the variation in a response variable y in terms of k predictor variables $x_1, ..., x_k$. We describe the mean value of y_i, the response for the ith individual, as

$$E(y_i|\beta, X) = \beta_1 x_{i1} + ... + \beta_k x_{ik}, i = 1, ..., n,$$

where $x_{i1}, ..., x_{ik}$ are the predictor values for the ith individual and $\beta_1, ..., \beta_k$ are unknown regression parameters. If we let $x_i = (x_{i1}, ..., x_{ik})$ denote the row vector of predictors for the ith individual and $\beta = (\beta_1, ..., \beta_k)$ the column vector of regression coefficients, we can reexpress the mean value as

$$E(y_i|\beta, X) = x_i\beta.$$

The $\{y_i\}$ are assumed to be conditionally independent given values of the parameters and the predictor variables. In the ordinary linear regression setting, we assume equal variances where $\text{var}(y_i|\theta, X) = \sigma^2$. We let $\theta = (\beta_1, ..., \beta_k, \sigma^2)$ denote the vector of unknown parameters. Finally, we assume that the errors $\epsilon_i = y_i - E(y_i|\beta, X)$ are independent normally distributed with mean 0 and variance σ^2.

In matrix notation, this model can be written for all observations as

$$y|\beta, \sigma^2, X \sim N_n(X\beta, \sigma^2 I),$$

where y is the vector of observations; X is the *design matrix* with rows $x_1, ..., x_n$; I is the identity matrix; and $N_k(\mu, A)$ indicates a multivariate normal distribution of dimension k with mean vector μ and variance-covariance matrix A.

To complete the Bayesian formulation of the model, we assume (β, σ^2) have the typical noninformative prior

$$g(\beta, \sigma^2) \propto \frac{1}{\sigma^2}.$$

9.2.2 The Posterior Distribution

The posterior analysis for the normal regression model has a form similar to the posterior analysis of a mean and variance for a normal sampling model. We represent the joint density of (β, σ^2) as the product

$$g(\beta, \sigma^2|y) = g(\beta|y, \sigma^2)g(\sigma^2|y).$$

The posterior distribution of the regression vector β conditional on the error variance σ^2, $g(\beta|y, \sigma^2)$, is multivariate normal with mean $\hat{\beta}$ and variance-covariance matrix $V_\beta \sigma^2$, where

$$\hat{\beta} = (X'X)^{-1}X'y, \ V_\beta = (X'X)^{-1}.$$

If one defines the inverse gamma(a, b) density proportional to $y^{-a-1}\exp\{-b/y\}$, then the marginal posterior distribution of σ^2 is inverse gamma$((n-k)/2, S/2)$, where

$$S = (y - X\hat{\beta})^T(y - X\hat{\beta}).$$

9.2.3 Prediction of Future Observations

Suppose we are interested in predicting a future observation $\tilde{y}$ corresponding to a covariate vector x^*. From the regression sampling model we have that $\tilde{y}$, conditional on β and σ^2, is $N(x^*\beta, \sigma)$. The posterior predictive density of $\tilde{y}$, $p(\tilde{y}|y)$, can be represented by a mixture of these sampling densities $p(\tilde{y}|\beta, \sigma^2)$, where they are averaged over the posterior distribution of the parameters β and σ^2:

$$p(\tilde{y}|y) = \int p(\tilde{y}|\beta, \sigma^2)g(\beta, \sigma^2|y)d\beta d\sigma^2.$$

9.2.4 Computation

The expressions for the posterior and predictive distributions lead to efficient simulation algorithms. To simulate from the joint posterior distribution of the regression coefficient vector β and the error variance σ^2, one

- simulates a value of the error variance σ^2 from its marginal posterior density $g(\sigma^2|y)$
- simulates a value of β from the conditional posterior density $g(\beta|\sigma^2, y)$.

Since the two component distributions (inverse gamma and multivariate normal) are convenient functional forms, it is relatively easy to construct an algorithm in R such as the one programmed in the function `blinreg` to perform this simulation.

Once the joint posterior distribution has been simulated, it is straightforward to obtain a sample from the marginal posterior distribution of any function $h(\beta, \sigma)$ of interest. For example, if x^* denotes a row vector of particular values of covariates, suppose one is interested in the mean response at x^*

$$E(y|x^*) = x^*\beta.$$

If β^* is a simulated draw from the marginal posterior of β, then $x^*\beta^*$ will be a simulated draw from the marginal posterior of $x^*\beta$. The R function `blinregexpected` facilitates the simulation of linear combinations of the beta coefficients.

Likewise the representation of the posterior predictive distribution of future response values suggests a simple algorithm for simulation. Suppose $\tilde{y}$ is a future response value corresponding to the row vector of covariates x^*. One simulates a single value of $\tilde{y}$ by

- simulating (β, σ^2) from the joint posterior given the data y
- simulating $\tilde{y}$ from its sampling density given the simulated values of β and σ^2

$$\tilde{y} \sim N(x^*\beta, \sigma).$$

The R function `blinregpred` can be used to simulate sets of draws of future observations corresponding to a list of covariate values of interest.

9.2.5 Model Checking

One method of assessing the goodness of fit of the model uses the posterior predictive distribution defined in the previous section. Suppose one simulates many samples $\tilde{y}_1, ..., \tilde{y}_n$ from the posterior predictive distribution conditional on the same covariate vectors $x_1, ..., x_n$ used to simulate the data. To judge if a particular response value y_i is consistent with the fitted model, one looks at the position of y_i relative to the histogram of simulated values of $\tilde{y}_i$ from the corresponding predictive distribution. If y_i is in the tail of the distribution, that indicates that this observation is a potential outlier.

A second approach is based on the use of "Bayesian residuals." In a traditional regression analysis, one judges the adequacy of the fitted model by inspection of the standardized residuals

$$r_i = \frac{y_i - x_i\hat{\beta}}{\hat{\sigma}\sqrt{1 - h_{ii}}},$$

where $\hat{\beta}$ and $\hat{\sigma}$ are the usual estimates of the regression vector and error standard deviation, and h_{ii} is the ith diagonal element of the "hat" matrix. From a Bayesian perspective, one can consider the distribution of the parametric residuals

$$\{\epsilon_i = y_i - x_i\beta\}.$$

Before any data are observed, the parametric residuals are a random sample from an $N(0, \sigma)$ distribution. Suppose we say that the ith observation is an outlier if $|\epsilon_i| > k\sigma$, where k is a predetermined constant such as 2 or 3. The prior probability that a particular observation is an outlier is $2\Phi(-k)$, where $\Phi(z)$ is the standard normal cdf.

After data y are observed, we can compute the posterior probability that each observation is an outlier. Define the functions z_1 and z_2 as

$$z_1 = (k - \hat{\epsilon}_i/\sigma)/\sqrt{h_{ii}}, \; z_2 = (-k - \hat{\epsilon}_i/\sigma)/\sqrt{h_{ii}},$$

where

$$\hat{\epsilon}_i = y_i - x_i\hat{\beta}.$$

Then the posterior probability that the ith observation is an outlier is

$$p_i = P(|\epsilon_i| > k\sigma|y) = \int (1 - \Phi(z_1) + \Phi(z_2))g(\sigma^2|y)d\sigma^2.$$

In practice, the p_is can be computed and compared to the prior probability $2\Phi(-k)$. The R function `bayesresiduals` can be used to compute the posterior outlying probabilities for a linear regression model.

9.2.6 An Example

Ramsey and Schafer (1997), chapter 10, describe an interesting study from Pimm et al (1988) on the extinction of birds. Measurements on breeding pairs of land-bird species were collected from 16 islands around Britain over the course of several decades. For each species, the dataset contains TIME, the average time of extinction on the islands where it appeared, NESTING, the average number of nesting pairs, SIZE, the size of the species (large or small), and STATUS, the migratory status of the species (migrant or resident). The objective is to fit a model that describes the variation in the time of extinction of the bird species in terms of the covariates NESTING, SIZE, and STATUS.

This dataset is available as `birdextinct` in the LearnBayes package. We read in the datafile and construct some initial graphs. Since the TIME variable is strongly right-skewed, we initially transform it by a logarithm creating the variable LOGTIME. Fig. 9.1, Fig. 9.2, and Fig. 9.3 plot LOGTIME against each of the three predictor variables. Since the categorical variables SIZE and STATUS take only two values, we use the R `jitter` function to jitter the horizontal location of the points so we can see any overlapping points. Note that there is a positive relationship between the average number of nesting pairs and time to extinction. However, there are five particular species (labeled in the graph) with points that seem to vary from the general pattern. There may be relationships of each of the categorical variables with LOGTIME, but the strength of the relationship seems weak in comparison with the relationship of NESTING and LOGTIME.

```
> data(birdextinct)
> attach(birdextinct)
> logtime=log(time)
> plot(nesting,logtime)
> identify(nesting,logtime,label=species,n=5)
> plot(jitter(size),logtime,xaxp=c(0,1,1))
> plot(jitter(status),logtime,xaxp=c(0,1,1))
```

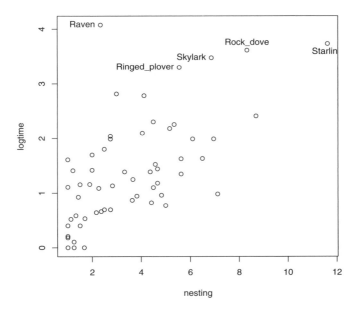

Fig. 9.1. Plot of logarithm of the extinction time against the average number of nesting pairs for the bird study.

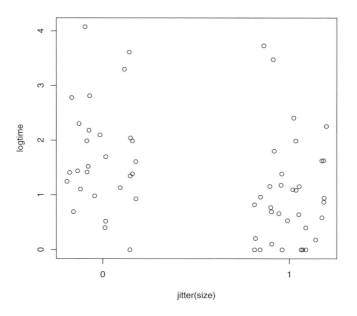

Fig. 9.2. Plot of the logarithm of the extinction time against the bird size for the bird study. The bird size variable is coded 0 for small and 1 for large.

We write the regression model as

$$E(\log TIME_i | x, \theta) = \beta_0 + \beta_1 NESTING_i + \beta_2 SIZE_i + \beta_3 STATUS_i.$$

As two of the covariates are categorical with two levels, they can be represented by binary indicators; in the data file **birdextinct**, SIZE is coded 0 (1) for small (large), and STATUS is coded 0 (1) for migrant (resident).

We first perform the traditional least-squares fit by the **lm** command.

```
> fit=lm(logtime~nesting+size+status,data=birdextinct,x=TRUE,y=TRUE)
> summary(fit)

Residuals:
    Min      1Q  Median      3Q     Max
-1.8410 -0.2932 -0.0709  0.2165  2.5167

Coefficients:
            Estimate Std. Error t value Pr(>|t|)
(Intercept)  0.43087    0.20706   2.081 0.041870 *
nesting      0.26501    0.03679   7.203 1.33e-09 ***
```

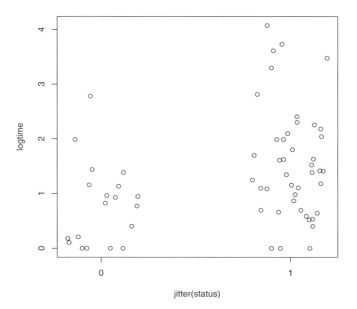

logtime

jitter(status)

Fig. 9.3. Plot of the logarithm of the extinction time against the bird status for the bird study. The bird status variable is coded 0 for migrant and 1 for resident.

```
size          -0.65220     0.16667   -3.913 0.000242 ***
status         0.50417     0.18263    2.761 0.007712 **
```

We see from the output that NESTING is a strong effect; species with larger number of nesting pairs tend to have longer extinction times which means that these species are less likely to be extinct. The SIZE and STATUS effects appear to be less significant; larger birds (with SIZE = 1) have smaller extinction times and resident birds (with STATUS = 1) have longer extinction times.

The function `blinreg` is used to sample from the joint posterior distribution of β and σ. The inputs to this function are the vector of values of the response variable y, the design matrix of the linear regression fit X, and the number of simulations m. Note that we used the optional arguments `x = TRUE, y = TRUE` in the function `lm` so that the design matrix and response vector are available as components of the structure `fit`.

```
> theta.sample=blinreg(fit$y,fit$x,5000)
```

The algorithm in `binreg` is based on the decomposition of the joint posterior $[\beta, \sigma^2|y]$ as the product $[\sigma^2|y][\beta|\sigma^2, y]$. To simulate one draw of (σ^2, β), σ^2 is first drawn from the inverse gamma$((n-k)/2, S/2)$ density:

```
S=sum(fit$residual^2)
```

```
shape=fit$df.residual/2; rate=S/2
sigma2=rigamma(1,shape,rate)
```

Then the regression vector β is simulated from the multivariate normal density with mean $\hat{\beta}$ and variance-covariance matrix $V_\beta \sigma^2$. Note that we obtain the matrix V_β by dividing the estimated variance-covariance matrix vcov from the least-squares fit by the mean square error stored in the variable MSE.

```
MSE = sum(fit$residuals^2)/fit$df.residual
vbeta=vcov(fit)/MSE
beta=rmnorm(1,mean=fit$coef,varcov=vbeta*sigma2)
```

The function blinreg returns two components, beta is a matrix of simulated draws from the marginal posterior of β, where each row is a simulated draw, and sigma is a vector of simulated draws from the marginal posterior of σ.

The following R commands construct histograms of the simulated posterior draws of the individual regression coefficients β_1, β_2, and β_3 and the error standard deviation σ (see Fig. 9.4).

```
> par(mfrow=c(2,2))
> hist(theta.sample$beta[,2],main="NESTING",
+   xlab=expression(beta[1]))
> hist(theta.sample$beta[,3],main="SIZE",
+   xlab=expression(beta[2]))
> hist(theta.sample$beta[,4],main="STATUS",
+   xlab=expression(beta[3]))
> hist(theta.sample$sigma,main="ERROR SD",
+   xlab=expression(sigma))
```

We can summarize each individual parameter by computing the 5th, 50th, and 95th percentiles of each collection of simulated draws. In the output, we use the apply and quantile commands to summarize the simulation matrix of β theta.sample$beta. Similarly, we use the quantile command to simulate the draws of σ.

```
> apply(theta.sample$beta,2,quantile,c(.05,.5,.95))
```

	X(Intercept)	Xnesting	Xsize	Xstatus
5%	0.09789072	0.2038980	-0.9374168	0.2050562
50%	0.42705148	0.2648745	-0.6475561	0.5024234
95%	0.77067086	0.3259122	-0.3803261	0.8082491

```
> quantile(theta.sample$sigma,c(.05,.5,.95))
```

5%	50%	95%
0.5679346	0.6576295	0.7725279

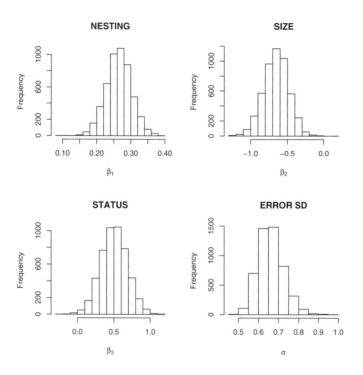

Fig. 9.4. Histogram of simulated draws from the marginal posterior distributions of β_1, β_2, β_3, and σ.

As expected, the posterior medians of the regression parameters are similar in value to the ordinary regression estimates. Actually they are equivalent since we applied a vague prior for β; any small differences between the posterior medians and the least-square estimates are due to small errors inherent in the simulation.

Next, suppose we are interested in estimating the mean log extinction time $E(y|x^*) = x^*\beta$ for four nesting pairs and for different combinations of SIZE and STATUS. The values of the four sets of covariates are shown in Table 9.1.

Table 9.1. Four sets of covariates of interest in the bird study.

Covariate Set	Nesting Pairs	Size	Status
A	4	small	migrant
B	4	small	resident
C	4	large	migrant
D	4	large	resident

In the following input, we define the four sets of covariates and stack these sets in the matrix X1. The function `blinregexpected` will give a simulated sample for the expected response $E(y|x^*) = x^*\beta$ for each set of covariate values. The inputs to the function are the matrix X1 of covariate values and the list of simulated values of β and σ obtained from the function `binlinreg`. The output of the function is a matrix where a column contains the simulated draws for a given covariate set. We construct histograms of the simulated draws for each of the mean extinction times and the plots are displayed in Fig. 9.5.

```
> cov1=c(1,4,0,0)
> cov2=c(1,4,1,0)
> cov3=c(1,4,0,1)
> cov4=c(1,4,1,1)
> X1=rbind(cov1,cov2,cov3,cov4)
> mean.draws=blinregexpected(X1,theta.sample)
> par(mfrow=c(2,2))
> hist(mean.draws[,1],main="Covariate set A",xlab="log TIME")
> hist(mean.draws[,2],main="Covariate set B",xlab="log TIME")
> hist(mean.draws[,3],main="Covariate set C",xlab="log TIME")
> hist(mean.draws[,4],main="Covariate set D",xlab="log TIME")
```

In the preceding work, we were interested in learning about the mean response value $E(y|x^*)$ for a given set of covariate values. Instead, suppose we are interested in predicting a future response $\tilde{y}$ for a given covariate vector x^*. The function `blinregpred` will produce a simulated sample of future response values for a regression model. Similar to the function `binlinregexpected`, the inputs to the function `blinregpred` are a matrix X1 where each row corresponds to a covariate set and the structure of simulated values of the parameters β and σ.

```
> cov1=c(1,4,0,0)
> cov2=c(1,4,1,0)
> cov3=c(1,4,0,1)
> cov4=c(1,4,1,1)
> X1=rbind(cov1,cov2,cov3,cov4)
> pred.draws=blinregpred(X1,theta.sample)
> par(mfrow=c(2,2))
> hist(pred.draws[,1],main="Covariate set A",xlab="log TIME")
> hist(pred.draws[,2],main="Covariate set B",xlab="log TIME")
> hist(pred.draws[,3],main="Covariate set C",xlab="log TIME")
> hist(pred.draws[,4],main="Covariate set D",xlab="log TIME")
```

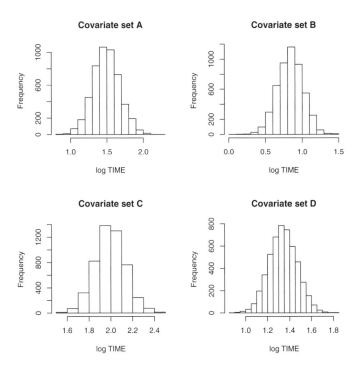

Fig. 9.5. Histograms of simulated draws of the posterior of the mean extinction time for four sets of covariate values.

Fig. 9.6 displays histograms of the simulated draws from the predictive distribution for the same four sets of covariates. Comparing Fig. 9.5 and Fig. 9.6 , note that the predictive distributions are substantially wider than the mean response distributions.

We illustrate two methods of checking if the observations are consistent with the fitted model. The first method is based on the use of the posterior predictive distribution described in Section 9.2.5. Let y_i^* denote the density of a future log extinction time for a bird with covariate vector x_i. Using the function `binregpred` we can simulate draws of the posterior predictive distributions for all $y_1^*, ..., y_{62}^*$ by using `fit$x` as an argument. In the R code, we summarize each predictive distribution by the 5th and 95th quantiles and graph these distributions as line plots using the `matplot` command (see Fig. 9.7). We place the actual log extinction times $y_1, ..., y_{62}$ as solid dots in the figure. We are looking to see if the observed response values are consistent with the corresponding predictive distributions; any points that fall outside of the corresponding 90% interval band are possible outliers. There are three points that exceed the 95th percentile that correspond to the species snipe, raven, and skylark.

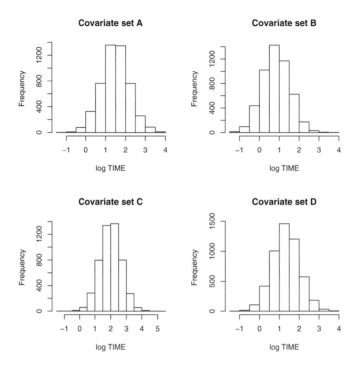

Fig. 9.6. Histograms of simulated draws of the predictive distribution for a future extinction time for four sets of covariate values.

```
> pred.draws=blinregpred(fit$x,theta.sample)
> pred.sum=apply(pred.draws,2,quantile,c(.05,.95))
> par(mfrow=c(1,1))
> ind=1:length(logtime)
> matplot(rbind(ind,ind),pred.sum,type="l",lty=1,col=1,
+   xlab="INDEX",ylab="log TIME")
> points(ind,logtime,pch=19)
```

Another method for outlier detection is based on the use of the Bayesian residuals $\epsilon_i = y_i - x_i\beta$. Following the strategy described in Section 9.2.5, we can compute the posterior outlying probabilities

$$P(|\epsilon_i| > k|y),$$

for all observations for a constant value k. These probabilities can be computed using the function `bayesresiduals`. The inputs are the `lm` fit structure `fit`, the matrix of simulated parameter draws `theta.sample`, and the value of k. The output is a vector of posterior outlying probabilities. In this example,

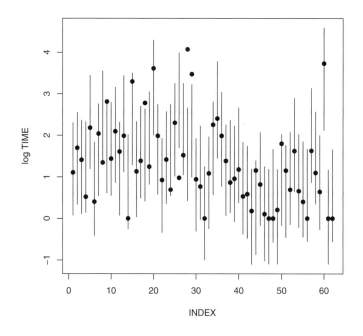

Fig. 9.7. Posterior predictive distributions of $\{y_i^*\}$ with actual log extinction times $\{y_i\}$ indicated by solid points.

we use a cutoff value of $k = 2$. We use the `plot` command to construct a scatterplot of the probabilities against the nesting covariate; the resulting display is in Fig. 9.8. By use of the `identify` command, we identify four birds that have outlying probabilities of .4 or higher. These birds have extinction times that are not well-explained by the variables NESTING, SIZE, and STATUS. Two of the outlying species, raven and skylark, were also identified by the posterior predictive methodology.

```
> prob.out=bayesresiduals(fit,theta.sample,2)
> par(mfrow=c(1,1))
> plot(nesting,prob.out)
> identify(nesting,prob.out,label=species,n=4)
```

9.3 Survival Modeling

Suppose one is interested in constructing a model for lifetimes in a survival study. For a set of n individuals, one observes the lifetimes $t_1, ..., t_n$. It is

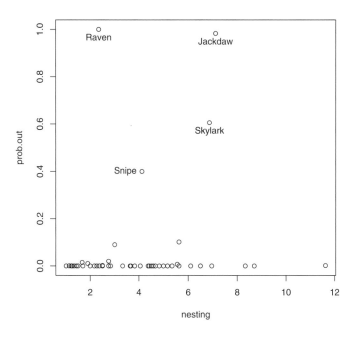

Fig. 9.8. Plot of posterior probabilities of outlying for all observations. Four unusually large probabilities are identified with the name of the species.

possible that some of the lifetimes are not observable since some individuals are still alive at the end of the study. In this case we represent the response by the pair (t_i, δ_i), where t_i is the observation and δ_i is a censoring indicator. If $\delta_i = 1$, the observation is not censored and t_i is the actual survival time. Otherwise when $\delta_i = 0$, the observation t_i is the censored time.

Suppose we wish to describe the variation in the survival times using p covariates $x_1, ..., x_p$. One can describe this relationship by the Weibull proportional hazards model. This model can be expressed as the log-linear model

$$\log t_i = \mu + \beta_1 x_{i1} + ... + \beta_p x_{ip} + \sigma \epsilon_i,$$

where $x_{i1}, ..., x_{ip}$ are the values of the p covariates for the ith individual and ϵ_i is assumed to have a Gumbel distribution with density $f(\epsilon) = \exp(\epsilon - e^\epsilon)$. There are $p+2$ unknown parameters in this model, the p regression coefficients, the constant term μ, and the scale parameter σ.

It can be shown that the density of the log time, $y_i = \log t_i$ is given by

$$f_i(y_i) = \frac{1}{\sigma} \exp(z_i - e^{z_i}),$$

where $z_i = (y_i - \mu - \beta_1 x_{i1} - ... - \beta_p x_{ip})/\sigma$. Also, the survival function for the ith individual is given by $S_i(y_i) = \exp(-e^{z_i})$. Then the likelihood function of the regression vector $\beta = (\beta_1, ..., \beta_p)$, μ and σ is given by

$$L(\beta, \mu, \sigma) = \prod_{i=1}^{n} \{f_i(y_i)\}^{\delta_i} \{S_i(y_i)\}^{1-\delta_i}.$$

Suppose we assign μ, β uniform priors and the scale parameter σ the usual noninformative prior proportional to $1/\sigma$. Then the posterior density is given, up to a proportionality constant, by

$$g(\beta, \mu, \sigma | \text{data}) \propto \frac{1}{\sigma} L(\beta, \mu, \sigma).$$

To illustrate the application of this model, Edmunson et al (1979) studied the effect of different chemotherapy treatments following surgical treatment of ovarian cancer. The response variable $TIME$ was the survival time in days following randomization to one of two chemotherapy treatments. Also we record a censoring variable $STATUS$ that indicates if $TIME$ is an actual survival time ($STATUS = 1$) or censored at that time ($STATUS = 0$). The two covariates are $TREAT$, the treatment group, and AGE, the age of the patient. The log-linear model is

$$\log TIME_i = \mu + \beta_1 TREAT_i + \beta_2 AGE_i + \sigma \epsilon_i.$$

The dataset is given the name **chemotherapy** in the LearnBayes package. To begin, we read in the dataset and illustrate fitting this model using the survreg function in the **survival** library.

```
> data(chemotherapy)
> attach(chemotherapy)
> library(survival)
> survreg(Surv(time,status)~factor(treat)+age,dist="weibull")

Call:
survreg(formula = Surv(time, status) ~ factor(treat) + age,
dist = "weibull")

Coefficients:
    (Intercept) factor(treat)2            age
    10.98683919     0.56145663    -0.07897718

Scale= 0.5489202

Loglik(model)= -88.7   Loglik(intercept only)= -98
        Chisq= 18.41 on 2 degrees of freedom, p= 1e-04
n= 26
```

Unlike the normal regression model, the posterior distribution of the parameters of this survival model can not be simulated by standard probability distributions. But we are able to apply our general computing strategy described in Chapter 6 to summarize the posterior distribution for this problem. We first make all parameters real-valued by transforming the scale parameter σ to $\eta = \log \sigma$. We write the following function weibullregpost, which computes the joint posterior density of $\theta = (\eta, \mu, \beta_1, \beta_2)$. The argument data is the data matrix where the first two columns are $\{t_i\}$ and $\{c_i\}$ and the remaining columns are the covariates TREAT and AGE.

```
weibullregpost=function(theta,data)
{
s=dim(data); k=s[2]; p=k-2
sp=dim(theta); N=sp[1]
t=data[,1]; c=data[,2]; X=data[,3:k]
sigma=exp(theta[,1])
mu=theta[,2]
beta=array(theta[,3:k],c(N,p))
n=length(t)
o=0*mu
for (i in 1:n)
{
  lp=0
  for (j in 1:p) lp=lp+beta[,j]*X[i,j]
  zi=(log(t[i])-mu-lp)/sigma
  fi=1/sigma*exp(zi-exp(zi))
  Si=exp(-exp(zi))
  o=o+c[i]*log(fi)+(1-c[i])*log(Si)
}
return(o)
}
```

To get some initial estimates at the location and spread of the posterior density, we use the laplace function. We use the output of the survreg fit to suggest the initial guess at the posterior mode $(-.5, 9, .5, -.05)$. The output of this function is the posterior mode $\hat{\theta}$ and associated variance-covariance matrix V.

```
> start=array(c(-.5,9,.5,-.05),c(1,4))
> d=cbind(time,status,treat-1,age)
> fit=laplace(weibullregpost,start,5,d)
> fit

$mode
           [,1]     [,2]       [,3]          [,4]
[1,] -0.5998159 10.98666 0.5614697 -0.07897431
```

```
$var
                [,1]            [,2]            [,3]             [,4]
[1,]    0.055357819   0.12113625    0.005351316   -0.0018662702
[2,]    0.121136245   1.62894124   -0.155833863   -0.0248862629
[3,]    0.005351316  -0.15583386    0.115557274    0.0017804199
[4,]   -0.001866270  -0.02488626    0.001780420    0.0003905468
```

```
$int
[1] -25.32347
```

We then use the information from the `laplace` function to find a proposal density for the Metropolis random walk chain programmed in the R function `rwmetrop`. The proposal density will be a multivariate normal density with mean 0 and variance-covariance scale V, where *scale* is a scale parameter chosen so that the random walk chain has an acceptance range in the 20–40% range. With some trial and error, we find that *scale* = 1.5 seems to give a satisfactory acceptance rate.

```
> proposal=list(var=fit$var,scale=1.5)
> bayesfit=rwmetrop(weibullregpost,proposal,fit$mode,10000,d)
> bayesfit$accept
```

```
[1] 0.2677
```

By use of several `hist` commands, we display histograms of the simulated draws from the marginal posterior densities of β_1 (corresponding to TREAT), β_2 (corresponding to AGE), and the scale parameter σ (see Fig. 9.9).

```
> par(mfrow=c(2,2))
> sigma=exp(bayesfit$par[,1])
> mu=bayesfit$par[,2]
> beta1=bayesfit$par[,3]
> beta2=bayesfit$par[,4]
> hist(beta1,xlab="treatment")
> hist(beta2,xlab="age",main="")
> hist(sigma,xlab="sigma",main="")
```

Suppose one is interested in estimating the survival curve for an individual in the treatment group (TREAT = 1) who is 60 years old. For a given time t, the probability that this individual survives beyond t days is given by

$$P(T > t) = \exp(-\exp(z)),$$

where $z = (\log t - \mu - \beta_1(1) - \beta_2(60))/\sigma$. A simulated sample of draws from this survival probability is obtained by computing this function on the simulated draws of θ, and this simulated sample can be summarized by the 5th, 50th, and

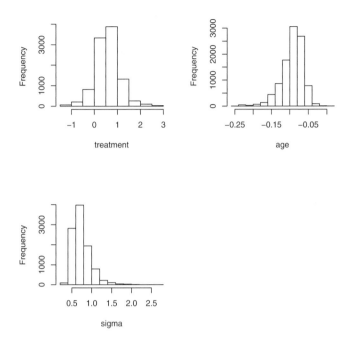

Fig. 9.9. Plot of the posterior probabilities of regression coefficients for TREAT and AGE and the scale parameter σ for the chemotherapy example.

95th percentiles. This procedure was repeated for a grid of t values between 0 and 2000 days. Fig. 9.10 graphs the 5th, 50th, and 95th percentiles for the survival curve for this individual. In a similar fashion, it is straightforward to make inferences about any function of the parameters of interest.

9.4 Further Reading

Chapter 14 of Gelman et al (2003) introduces Bayesian model building and inference for normal linear models. Analogous methods for generalized linear models are presented in chapter 16 of Gelman et al (2003). The Bayesian linear regression model is also described in chapter 6 of Gill (2002) and chapter 12 of Press (2003). The classical Weibull survival regression model is discussed in Collett (1994). Chaloner and Brant (1988) describes the use of Bayesian residuals in a linear regression model.

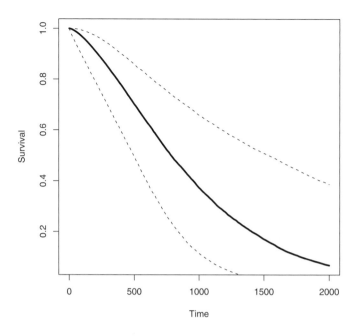

Fig. 9.10. Posterior median and 90% Bayesian interval estimates for the survival function S for an individual 60 years old in the treatment group.

9.5 Summary of R Functions

`bayesresiduals` – computation of posterior outlying probabilities for a linear regression model with a noninformative prior
Usage: `bayesresiduals(fit, theta.sample, k)`
Arguments: `fit`, output of a least-squares fit (R function `lm`); `theta.sample`, list with components `beta` (matrix of simulated draws from the posterior of beta) and `sigma` (vector of simulated draws from the posterior of sigma); `k`, cutoff value that defines an outlier
Value: vector of posterior outlying probabilities

`blinreg` – gives a simulated sample from the joint posterior distribution of the regression vector and the error standard deviation for a linear regression model with a noninformative prior
Usage: `blinreg(y,X,m)`
Arguments: `y`, vector of responses; `X`, design matrix; `m`, number of simulations desired
Value: `beta`, matrix of simulated draws of beta where each row corresponds to one draw; `sigma`, vector of simulated draws of the error standard deviation

blinregexpected – simulates draws of the expected response for a linear regression model with a noninformative prior
Usage: `binregexpected(X,theta.sample)`
Arguments: `X`, matrix where each row corresponds to a covariate set;
`theta.sample`, list with components `beta` (matrix of simulated draws from the posterior of beta) and `sigma` (vector of simulated draws from the posterior of sigma
Value: matrix where a column corresponds to the simulated draws of the expected response for a given covariate set

blinregpred - simulates draws of the predicted future response for a linear regression model with a noninformative prior
Usage: `binregpred(X,theta.sample)`
Arguments: `X`, matrix where each row corresponds to a covariate set;
`theta.sample`; list with components `beta` (matrix of simulated draws from the posterior of beta) and `sigma` (vector of simulated draws from the posterior of sigma
Value: matrix where a column corresponds to the simulated draws of the predicted future response for a given covariate set

weibullregpost – computes the logarithm of the posterior of (log sigma, mu, beta) for a Weibull proportional odds model
Usage: `weibullregpost(theta,data)`
Arguments: `theta`, matrix of parameter values where each row represents a value of (log sigma, mu, beta); `data`, matrix with columns survival time, censoring variable, and covariate matrix
Value: vector of values of the log posterior where each value corresponds to each row of the parameters in theta

9.6 Exercises

1. **Normal linear regression**
   Dobson (2001) describes a birthweight regression study. One is interested in predicting a baby's birthweight (in grams) based on the gestational age (in weeks) and the gender of the baby. The data are presented in Table 9.2 and available as `birthweight` in the LearnBayes package. In the standard linear regression model, we assume that

$$BIRTHWEIGHT_i = \beta_0 + \beta_1 AGE_i + GENDER_i + \epsilon,$$

   where the ϵ_i are independent and normally distributed with mean 0 and variance σ^2.

   a) Use the R function `lm` to fit this model by least-squares. From the output, assess if the effects AGE and GENDER are significant, and if they are significant, describe the effects of each covariate on birthweight.

Table 9.2. Birthweight (in grams) and gestational age (weeks) for male and female babies.

Male		Female	
Age	Birthweight	Age	Birthweight
40	2968	40	3317
38	2795	36	2729
40	3163	40	2935
35	2925	38	2754
36	2625	42	3210
37	2847	39	2817
41	3292	36	3126
40	3473	37	2539
37	2628	36	2412
38	3176	38	2991
40	3421	39	2875
38	2975	40	3231

b) Suppose a uniform prior is placed on the regression parameter vector $\beta = (\beta_0, \beta_1, \beta_2)$. Use the function `blinreg` to simulate a sample of 5000 draws from the joint posterior distribution of (β, σ^2). From the simulated sample, compute posterior means and standard deviations of β_1 and β_2. Check the consistency of the posterior means and standard deviations with the least-squares estimates and associated standard errors from the `lm` run.

c) Suppose one is interested in estimating the expected birthweight for male and female babies of gestational weeks 36 and 40. From the simulated draws of the posterior distribution and function `binregexpected`, construct 90% interval estimates for 36-week males, 36-week females, 40-week males, and 40-week females.

d) Suppose instead one wishes to predict the birthweight for a 36-week male, a 36-week female, a 40-week male, and a 40-week female. Use the function `blinregpred` and the simulated posterior sample to construct 90% prediction intervals for the birthweight for each type of baby.

2. **Logistic regression**
   For a given professional athlete, his or her performance level will tend to increase until midcareer and then deteriorate until retirement. Let y_i denote the number of home runs hit by the professional baseball player Mike Schmidt in n_i at-bats (opportunities) during the ith season. Table 9.3 gives Schmidt's age, y_i and n_i for all 18 years of his baseball career. The datafile is named `schmidt` in the LearnBayes package. The home run rates $\{y_i/n_i\}$ are graphed against Schmidt's year in Fig. 9.11. If y_i is assumed to be binomial(n_i, p_i), where p_i denotes the probability of hitting a home run during the ith season, then a reasonable model for the $\{p_i\}$ is the logit quadratic model of the form

$$\log\left(\frac{p_i}{1 - p_i}\right) = \beta_0 + \beta_1 AGE_i + \beta_2 AGE_i^2,$$

where AGE_i is Schmidt's age during the ith season.

Table 9.3. Home run hitting data for baseball player Mike Schmidt.

Age	Home Runs	At-Bats	Age	Home Runs	At-Bats
23	1	34	32	31	354
24	18	367	33	35	514
25	36	568	34	40	534
26	38	562	35	36	528
27	38	584	36	33	549
28	38	544	37	37	552
29	21	513	38	35	522
30	45	541	39	12	390
31	48	548	40	6	148

a) Assume that the regression vector $\beta = (\beta_0, \beta_1, \beta_2)$ has a uniform non-informative prior. Write a short R function to compute the logarithm of the posterior density of β.

b) Use the function `laplace` to find the posterior mode and associated variance-covariance matrix of β.

c) Based on the output from `laplace`, use the function `rwmetrop` to simulate 5000 draws from the posterior distribution of β.

d) One would expect the fitted parabola to have a concave down shape where $\beta_2 < 0$. Use the simulation output from part (c) to find the posterior probability that the fitted curve is concave down.

3. **Logistic regression (continued)**

For this exercise, we assume that a simulated sample from the posterior distribution of the regression vector β has been obtained.

a) When evaluating a baseball player, one is interested in estimating the player's ability at his peak. One can show that if $\beta_3 < 0$, the peak value of the probability, on the logit scale, is given by

$$PEAK = \beta_1 - \frac{\beta_2}{4\beta_3}.$$

Compute a density estimate of the marginal posterior density of PEAK.

b) One is also interested in the age at which a player achieves his peak performance. From the quadratic model, the peak age can be shown to be equal to

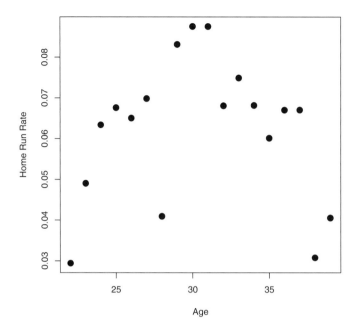

Fig. 9.11. Scatterplot of home run rates HR/AB against age for Mike Schmidt.

$$PEAK\ AGE = -\frac{\beta_2}{2\beta_2}.$$

Using the simulated draws from the posterior of β, find a 90% interval estimate for the PEAK AGE.

4. **Survival modeling**

Collett (1994) describes an investigation to evaluate a histochemical marker HPA, which discriminates between primary breast cancer that has metastasized and that which has not. The question is whether HPA staining can be used to predict the survival experience of women with breast cancer. Tumors of the women were treated with HPA, and each tumor was classified as being positively or negatively stained, positively staining corresponds to a tumor with the potential for metastasis. Survival times of the women who died of breast cancer were collected; the data are displayed in Table 9.4. For some women (indicated by an asterisk in Table 9.4), the survival status at the end of the study was unknown and the time from surgery to last date they were known to be alive is a censored survival time. The datafile **breastcancer** in the LearnBayes package contains the data. There are three variables: **time** is the survival time (in months); **status** gives the censoring status, where status $= 1$

indicates a complete survival time and status $= 0$ indicates a time that is censored; and stain indicates the group, where stain $= 0$ (1) indicates a tumor that was negatively (positively) stained.

Table 9.4. Survival times of women with tumors that were negatively or positively stained with HPA from Collett.

Negative Staining	Positive Staining	
23	5	68
47	5	71
69	10	78*
70*	13	105*
71*	18	107*
100*	24	109*
101*	26	113
148	26	116*
181	31	118
198*	35	143
208*	40	154*
212*	41	162*
224*	48	188*
	50	212*
	59	217*
	61	225*

a) Use the function survreg to fit a Weibull proportional hazards model of the form

$$\log TIME_i = \mu + \beta GROUP_i + \sigma \epsilon_i,$$

where ϵ_i is assumed to have a standard Gumbul distribution. Obtain estimates and associated standard errors for the group regression coefficient β and the scale parameter σ.

b) The function weibullregpost computes the log posterior of $(\log \sigma, \mu, \beta)$ assuming the standard noninformative prior. Use the function laplace to find the posterior mode and associated variance-covariance matrix. Then apply the function rwmetrop to simulate a sample of 1000 iterates from the joint posterior. Compute the posterior mean and standard deviation of β and σ and compare your answers with the estimates from part (a).

c) Using the simulated sample from the posterior of $(\log \sigma, \mu, \beta)$, estimate the survival curve $S(t)$ for a patient in the negatively stained group and a patient in the positively stained group. Choose a sequence of values of the time t, and for each t, find 5th, 50th, and 95th percentiles of the survival probability $S(t)$. As in Fig. 9.10, graph the median estimates of the survival curves for the two individuals.

10

Gibbs Sampling

10.1 Introduction

One attractive method for constructing an MCMC algorithm is Gibbs sampling, introduced in Chapter 6. To slightly generalize our earlier discussion, suppose that we partition the parameter vector of interest into p components $\theta = (\theta_1, ..., \theta_p)$, where θ_k may consist of a vector of parameters. The MCMC algorithm is implemented by sampling in turn from the p conditional posterior distributions

$$[\theta_1|\theta_2, ..., \theta_p], ..., [\theta_p|\theta_1, ..., \theta_{p-1}].$$

Under general regularity conditions, draws from this Gibbs sampler will converge to the target joint posterior distribution $[\theta_1, ..., \theta_p]$ of interest.

For a large group of inference problems, Gibbs sampling is automatic in the sense that all conditional posterior distributions are available or easy to simulate using standard probability distributions. There are several attractive aspects of "automatic" Gibbs sampling. First, one can program these simulation algorithms with a small amount of R code, especially when one can use vector and matrix representations for parameters and data. Second, unlike the more general Metropolis-Hastings algorithms described in Chapter 6, there are no tuning constants or proposal densities to define. Last, these Gibbs sampling algorithms provide a nice introduction to the use of more sophisticated MCMC algorithms in Bayesian fitting.

We illustrate the use of R to write Gibbs sampling algorithms for several popular inferential models. We revisit the robust modeling example of Section 6.8 where we applied various computational algorithms to summarize the exact posterior distribution. In Section 10.2, we illustrate a simple Gibbs sampler by representing the t sampling model as a scale mixture of normal densities. In Section 10.3, we apply the idea of latent variables to simulate from a binary response model where a probit link is used. This algorithm is attractive in that one can simulate from this probit model by iterating between truncated normal and multivariate normal probability distributions.

We conclude the chapter by considering a problem where one desires to smooth a two-way table of means. One model for these data is to assume that the underlying population means of the table follow a particular order restriction. A second model assumes that the population means follow a hierarchical regression model, where the population means are a linear function of row and column covariates. For both problems, R functions can be used to implement Gibbs sampling algorithms for simulating from the joint posterior of all parameters. These algorithms are automatic in that they are entirely based on standard probability distribution simulations.

10.2 Robust Modeling

We revisit the situation in Section 6.9 where we model data with a symmetric continuous distribution. When there is a possibility of outliers, a good strategy assumes the observations are distributed from a population with tails that are heavier than the normal form. One example of a heavy-tailed distribution is the t family with a small number of degrees of freedom.

With this motivation we suppose $y_1, ..., y_n$ are a sample from a t distribution with location μ, scale parameter σ, and known degrees of freedom ν. If we assign the usual noninformative prior on (μ, σ)

$$g(\mu, \sigma) \propto \frac{1}{\sigma},$$

the posterior density is given by

$$g(\mu, \sigma | y) \propto \frac{1}{\sigma} \prod_{i=1}^{n} \frac{1}{\sigma} \left(1 + \frac{(y_i - \mu)^2}{\sigma^2} \right)^{-(\nu+1)/2}.$$

In the case of Cauchy sampling ($\nu = 1$), we illustrated in Section 6.9 the use of different computational algorithms to summarize this representation of the posterior density.

By use of a simple trick, we can implement an automatic Gibbs sampler for this problem. A t density with location μ, scale σ, and degrees of freedom ν can be represented as the following mixture:

$$y|\lambda \sim N(\mu, \sigma/\sqrt{\lambda}), \ \lambda \sim \text{gamma}(\nu/2, \nu/2).$$

Suppose each observation y_i is represented as a scale mixture of normals with the introduction of the scale parameter λ_i. Then we can write our model as

$$y_i|\lambda_i \sim N(\mu, \sigma/\sqrt{\lambda_i}), \ i = 1, ..., n$$
$$\lambda_i \sim \text{gamma}(\nu/2, \nu/2), \ i = 1, ..., n$$
$$(\mu, \sigma) \sim g(\mu, \sigma) \propto 1/\sigma.$$

In the following, it is convenient to express the posterior in terms of the variance σ^2 instead of the standard deviation σ. Using the scale-mixture representation, the joint density of all parameters $(\mu, \sigma^2, \{\lambda_i\})$ is given by

$$\frac{1}{\sigma^2} \prod_{i=1}^{n} \left(\frac{\lambda_i^{1/2}}{\sigma} \exp\left[-\frac{\lambda_i}{2\sigma^2}(y_i - \mu)^2 \right] \right) \prod_{i=1}^{n} \left(\lambda_i^{\nu/2-1} \exp\left[-\frac{\nu\lambda_i}{2} \right] \right).$$

On the surface, it appears that we have complicated the analysis through the introduction of the scale parameters $\{\lambda_i\}$. But Gibbs sampling is easy now since all of the conditional distributions have the following simple functional forms:

1. Conditional on μ and σ^2, $\lambda_1, ..., \lambda_n$ are independent where

$$\lambda_i \sim \text{gamma}\left(\frac{\nu + 1}{2}, \frac{(y_i - \mu)^2}{2\sigma^2} + \frac{\nu}{2} \right).$$

2. Conditional on σ^2 and $\{\lambda_i\}$, the mean μ has a normal distribution:

$$\mu \sim N\left(\frac{\sum_{i=1}^{n} \lambda_i y_i}{\sum_{i=1}^{n} \lambda_i}, \frac{\sigma}{\sqrt{\sum_{i=1}^{n} \lambda_i}} \right).$$

3. Conditional on μ and $\{\lambda_i\}$, the variance σ^2 has an inverse gamma distribution:

$$\sigma^2 \sim \text{inv} - \text{gamma}\left(\frac{n}{2}, \frac{\sum_{i=1}^{n} \lambda_i (y_i - \mu)^2}{2} \right).$$

In R, we can let `lam` denote the vector $\{\lambda_i\}$, and `mu` and `sig2` denote the values of μ and σ^2. These three conditional distribution simulations can be implemented by the following R commands:

```
lam=rgamma(n,shape=(v+1)/2,rate=v/2+(y-mu)^2/2/sig2)
mu=rnorm(1,mean=sum(y*lam)/sum(lam),sd=sqrt(sig2/sum(lam)))
sig2=rigamma(1,n/2,sum(lam*(y-mu)^2)/2)
```

Note that we are using the random gamma function `rgamma` using a vector `rate` parameter; due to the conditional independence property, $\lambda_1, ..., \lambda_n$ can be simultaneously simulated by a single command. Also we have defined the function `rigamma` in the LearnBayes package to simulate from the inverse gamma density $y^{-a-1} \exp(-b/y)$ with arguments a and b.

The function `robustt` will implement this Gibbs sampling algorithm. The three arguments to this function are the data vector `y`, the degrees of freedom `v`, and the number of cycles of the Gibbs sampler `m`. The output of this function is a list with three components: `mu` is a vector of simulated draws of μ, `s2` is a vector of simulated draws of σ^2, and `lam` is a matrix of simulated draws of $\{\lambda_i\}$, where each row corresponds to a single draw.

We apply this algorithm to Darwin's dataset of the differences of the heights of cross- and self-fertilized plants analyzed in Chapter 6. We model the observations with a t(4) density and run the algorithm for 10,000 cycles.

```
> data(darwin)
> attach(darwin)
> fit=robustt(difference,4,10000)
```

We use the density estimation command `density` to construct a smooth estimate of the marginal posterior density of the location parameter μ. The resulting graph is shown in Fig. 10.1.

```
> plot(density(fit$mu),xlab="mu")
```

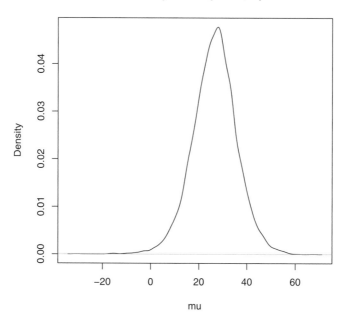

Fig. 10.1. Density estimate of simulated sample of marginal posterior density of μ in t modeling example.

The $\{\lambda_i\}$ parameters are interesting to examine since λ_i represents the weight of the observation y_i in the estimation of the location and scale parameters of the t population. In the following R code, we compute the posterior mean of each λ_i and place the posterior means in the vector `mean.lambda`. Likewise, we compute the 5th and 95th percentiles of each simulated sample of $\{\lambda_i\}$ (by use of the `apply` command with the function `quantile`) and store these quantiles in the vectors `lam5` and `lam95`. We first plot the posterior means of the $\{\lambda_i\}$ against the observations $\{y_i\}$, then we overlay lines that represent 90% interval estimates for these parameters (see Fig. 10.2). Note that the location of the posterior density of λ_i tends to be small for the outlying observations; these particular observations are downweighted in the estimation of the location and scale parameters.

```
> mean.lambda=apply(fit$lam,2,mean)
> lam5=apply(fit$lam,2,quantile,.05)
> lam95=apply(fit$lam,2,quantile,.95)
> plot(difference,mean.lambda,lwd=2,ylim=c(0,3),ylab="Lambda")
> for (i in 1:length(difference))
+    lines(c(1,1)*difference[i],c(lam5[i],lam95[i]))
> points(difference,0*difference-.05,pch=19,cex=2)
```

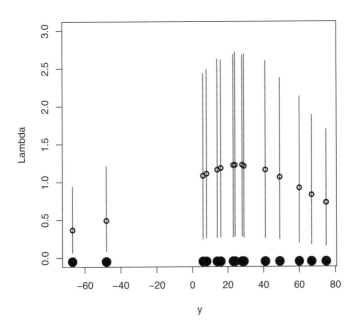

Fig. 10.2. 90% posterior interval estimates of scale parameters $\{\lambda_i\}$ plotted against the observations y. The observations are also plotted along the horizontal axis.

10.3 Binary Response Regression with a Probit Link

In Section 4.4, we considered a regression problem where we modeled the probability of death as a function of the dose level of a compound. We now consider the more general case where a probability is represented as a function of several covariates. By regarding this problem as a missing data problem, one can develop an automatic Gibbs sampling method described in Albert and Chib (1993) for simulating from the posterior distribution.

Suppose one observes binary observations $y_1, ..., y_n$. Associated with the ith response, one observes the values of k covariates $x_{i1}, ..., x_{ik}$. In the probit regression model, the probability that $y_i = 1$, p_i, is written as

$$p_i = P(y_i = 1) = \Phi(x_{i1}\beta_1 + ... + x_{ik}\beta_k),$$

where $\beta = (\beta_1, ..., \beta_k)$ is a vector of unknown regression coefficients and $\Phi()$ is the cdf of a standard normal distribution. If we place a uniform prior on β, then the posterior density is given by

$$g(\beta|y) \propto \prod_{i=1}^{n} p_i^{y_i} (1 - p_i)^{1-y_i}.$$

In the example to be discussed shortly, the binary response y_i is an indicator of survival, where $y_i = 1$ indicates the person survived the ordeal and $y_i = 0$ indicates the person did not survive. Suppose that there exists a continuous measurement Z_i of health such that if Z_i is positive, then the person survives; otherwise the person does not survive. Moreover the health measurement is related to the k covariates by the normal regression model

$$Z_i = x_{i1}\beta_1 + ... + x_{ik}\beta_k + \epsilon_i,$$

where $\epsilon_1, ..., \epsilon_n$ are a random sample from a standard normal distribution. It is a straightforward calculation to show that

$$P(y_i = 1) = P(Z_i > 0) = \Phi(x_{i1}\beta_1 + ... + x_{ik}\beta_k).$$

So we can regard this problem as a missing data problem where we have a normal regression model on latent data $Z_1, ..., Z_n$ and the observed responses are missing or incomplete in that we only observe if $Z_i > 0$ $(y_i = 1)$ or $Z_i \leq 0$ $(y_i = 0)$.

An automatic Gibbs sampling algorithm is constructed by adding the (unknown) latent data $Z = (Z_1, ..., Z_n)$ to the parameter vector β and sampling from the joint posterior distribution of Z and β. Both conditional posterior distributions, $[Z|\beta]$ and $[\beta|Z]$, have convenient functional forms. If we are given a value of the vector of latent data Z, then it can be shown that the conditional posterior distribution of β is

$$[\beta|Z, \text{data}] \sim N_k((X'X)^{-1}X'Z, (X'X)^{-1}),$$

where X is the design matrix for the problem. If we are given a value of the regression parameter vector β, then $Z_1, ..., Z_n$ are independent, with

$$[Z_i|\beta, \text{data}] \sim N(x_i\beta, 1)I(Z_i > 0), \text{ if } y_i = 1,$$

$$[Z_i|\beta, \text{data}] \sim N(x_i\beta, 1)I(Z_i < 0), \text{ if } y_i = 0,$$

and x_i denotes the vector of covariates for the ith individual. So given the value of β, we simulate the latent data Z from truncated normal distributions, where the truncation point is 0 and the side of the truncation depends on the values of the binary response.

The function `bayes.probit` implements this Gibbs sampling algorithm for the probit regression model. The key lines in the R code of this function simulate from the two conditional distributions. To simulate a variate Z from a normal($\mu, 1$) distribution truncated on the interval (a, b), one uses the recipe

$$Z = \Phi^{-1}\left[\Phi(a - \mu) + U(\Phi(b - \mu) - \Phi(a - \mu))\right] + \mu,$$

where $\Phi()$ and $\Phi^{-1}()$ are, respectively, the standard normal cdf and inverse cdf, and U is a uniform variate on the unit interval. In the following code, `lp` is the vector of linear predictors and `y` is the vector of binary responses. Then the latent data `z` are simulated by the following code:

```
lp=x%*%beta
bb=pnorm(-lp)
tt=(bb*(1-y)+(1-bb)*y)*runif(n)+bb*y
z=qnorm(tt)+lp
```

Given values of the latent data in the vector `z` and the design matrix in `x`, the following code simulates the vector data from the multivariate normal distribution:

```
v=solve(t(x)%*%x)
mn=solve(t(x)%*%x)%*%(t(x)%*%z)
beta=rmnorm(1,mean=c(mn),varcov=v)
```

To illustrate the use of the function `bayes.probit`, we consider a dataset on the Donner party, a group of wagon train emigrants who had difficulty in crossing the Sierra Nevada mountains in California and a large number starved to death. (See Grayson (1990) for more information about the Donner party.) The dataset `donner` in the LearnBayes package contains the age, gender, and survival status for 45 members of the party age 15 and older. For the ith member, we let y_i denote the survival status (1 if survived, 0 if not survived), $MALE_i$ denote the gender (1 if male, 0 if female), and AGE_i denote the age in years. We wish to fit the model

$$P(y_i = 1) = \Phi(\beta_0 + \beta_1 MALE_i + \beta_2 AGE_i).$$

We read in the dataset that has variable names `survival`, `male`, and `age`. We create the design matrix and store it in the variable `X`.

```
> data(donner)
> attach(donner)
> X=cbind(1,age,male)
```

A maximum likelihood fit of the probit model can be found using the `glm` function with the `family=binomial` option, indicating by `link=probit` that a probit link is used.

```
> fit=glm(survival~X-1,family=binomial(link=probit))
> summary(fit)
```

```
Call:
glm(formula = survival ~ X - 1, family = binomial(link = probit))
```

```
Coefficients:
      Estimate Std. Error z value Pr(>|z|)
X      1.91730    0.76438   2.508   0.0121 *
Xage  -0.04571    0.02076  -2.202   0.0277 *
Xmale -0.95828    0.43983  -2.179   0.0293 *
---
Signif. codes:  0 '***' 0.001 '**' 0.01 '*' 0.05 '.' 0.1 ' ' 1
```

To fit the posterior distribution of β by Gibbs sampling, we use the function `bayes.probit`. The inputs to this function are the vector of binary responses `survival`, the design matrix `X`, and the number of cycles of Gibbs sampling `m`.

```
> m=10000
> fit=bayes.probit(survival,X,m)
```

The output of this function is a matrix of simulated draws, where each row corresponds to a single draw of β. We can compute the posterior means and posterior standard deviations of the regression coefficients by use of the `apply` function.

```
> apply(fit,2,mean)
```

```
[1]   2.10178712 -0.05090274 -1.00917397
```

```
> apply(fit,2,sd)
```

```
[1] 0.78992508 0.02127450 0.45329737
```

The posterior mean and standard deviations are similar in value to the maximum likelihood estimates and their associated standard errors. This is expected since the posterior analysis was based on a noninformative prior on the regression vector β.

Since both the age and gender variables appear to be significant in this study, it is interesting to explore the probability of survival

$$p = P(y = 1) = \Phi(\beta_0 + \beta_1 AGE + \beta_2 MALE)$$

as a function of these two variables. The function `bprobit.probs` is useful for computing a simulated posterior sample of probabilities for covariate sets of interest. For example, suppose we wish to estimate the probability of survival for males age 15 through 65. We construct a matrix of covariate vectors `X1`, where a row corresponds to the values of the covariates for a male of a particular age. The function `bprobit.probs` is used with inputs `X1` and the simulated matrix of simulated regression coefficients from `bayes.probit`. The output is a matrix of simulated draws `p.male`, where each column corresponds to a simulated sample for a given survival probability.

```
> a=seq(15,65)
> X1=cbind(1,a,1)
> p.male=bprobit.probs(X1,fit)
```

We can summarize the simulated matrix of probabilities by the `apply` command. We compute the 5th, 25th, and 95th percentiles of the simulated sample of

$$p = \Phi(\beta_0 + \beta_1 AGE + \beta_2(MALE = 1))$$

for each of the AGE values. In Fig. 10.3, we graph these percentiles as a function of age. For each age, the solid line is the location of the median of the survival probability and the interval between the dashed lines corresponds to a 90% interval estimate for this probability. In Fig. 10.4, we repeat this work to estimate the survival probabilities of females of different ages. These two figures clearly show how survival is dependent on the age and gender of the emigrant.

```
> plot(a,apply(p.male,2,quantile,.5),type="l",ylim=c(0,1),
+    xlab="age",ylab="Probability of Survival")
> lines(a,apply(p.male,2,quantile,.05),lty=2)
> lines(a,apply(p.male,2,quantile,.95),lty=2)
```

10.4 Estimating a Table of Means

10.4.1 Introduction

A university would like its students to be successful in their classes. Since all students do not do well and some may eventually drop out, the admissions office is interested in understanding what measures of high school performance are helpful in predicting success in college. The standard measure of performance in university courses is the grade point average (GPA). The admissions

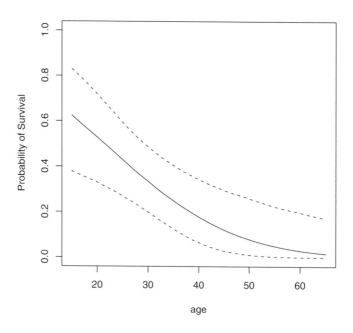

Fig. 10.3. Posterior distribution of probability of survival for males of different ages. For each age, the 5th, 50th, and 95th percentiles of the posterior are plotted.

people are interested in understanding the relationship between a student's GPA with two particular high school measures: the student's score on the ACT exam (a standardized test given to all high school juniors) and the student's percentile rank in his or her high school class.

The datafile `iowagpa` in the LearnBayes package contains the data for this problem. This dataset is a matrix of 40 rows, where a row contains the sample mean, the sample size, the high school rank percentile and the ACT score. By use of the R `matrix` command, these data are represented by the following two-way table of means. The row of the table corresponds to the high school rank (HSR) of the student and the column corresponds to the level of the ACT score. The entry of the table is the mean GPA of all students with the particular high school rank and ACT score.

```
> data(iowagpa)
> rlabels = c("91-99", "81-90", "71-80", "61-70", "51-60",
  "41-50", +      "31-40", "21-30")
> clabels = c("16-18", "19-21", "22-24", "25-27", "28-30")
> gpa = matrix(iowagpa[, 1], nrow = 8, ncol = 5, byrow = T)
```

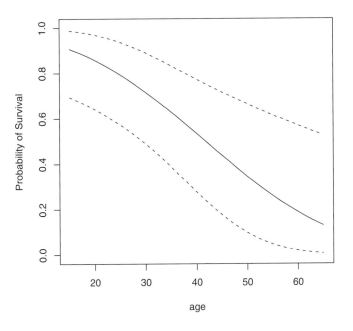

age

Fig. 10.4. Posterior distribution of probability of survival for females of different ages. For each age, the 5th, 50th, and 95th percentiles of the posterior are plotted.

```
> dimnames(gpa) = list(HSR = rlabels, ACTC = clabels)
> gpa
```

	ACTC				
HSR	16-18	19-21	22-24	25-27	28-30
91-99	2.64	3.10	3.01	3.07	3.34
81-90	2.24	2.63	2.74	2.76	2.91
71-80	2.43	2.47	2.64	2.73	2.47
61-70	2.31	2.37	2.32	2.24	2.31
51-60	2.04	2.20	2.01	2.43	2.38
41-50	1.88	1.82	1.84	2.12	2.05
31-40	1.86	2.28	1.67	1.89	1.79
21-30	1.70	1.65	1.51	1.67	2.33

The following table gives the number of students in each level of high school rank and ACT score. Note that most of the students are in the upper-right corner of the table corresponding to high values of both variables.

```
> samplesizes = matrix(iowagpa[, 2], nrow = 8, ncol = 5, byrow = T)
> dimnames(samplesizes) = list(HSR = rlabels, ACTC = clabels)
> samplesizes
```

HSR	ACTC				
	16-18	19-21	22-24	25-27	28-30
91-99	8	15	78	182	166
81-90	20	71	168	178	91
71-80	40	116	180	133	46
61-70	34	93	124	101	19
51-60	41	73	62	58	9
41-50	19	25	36	49	16
31-40	8	9	15	29	9
21-30	4	5	9	11	1

The admissions people at this university believe that both high school rank and ACT score are useful predictors of grade point average. One way of expressing this belief is to state that the corresponding population means of the table satisfy a particular order restriction. Let μ_{ij} denote the mean GPA of the population of students with the ith level of HSR and jth level of ACT score. If one looks at the ith row of the table with a fixed HSR rank, it is reasonable to believe that the column means satisfy the order restriction

$$\mu_{i1} \leq \mu_{i2} \leq \cdots \leq \mu_{i5}.$$

This expresses the belief that if you focus on students with a given high-school rank, then students with higher ACT scores will obtain higher grade point averages. Likewise, for a particular ACT level (jth column), one may believe that students with higher percentile ranks will get higher grades, and thus the row means satisfy the order restriction

$$\mu_{1j} \leq \mu_{2j} \leq \cdots \leq \mu_{9j}.$$

The standard estimates of the population means are the corresponding observed sample means. Fig. 10.5 displays the matrix of sample means using a series of line graphs where each row of means is represented by a single line. (This graph is created using the R function matplot.) Note from the figure that the sample means do not totally satisfy the order restrictions. For example, in the "31–40" row of HSR, the mean GPA for ACT score 19–21 is larger than the mean GPA in the same row for larger values of ACT. It is desirable to obtain smoothed estimates of the population means that more closely follow the belief in order restriction.

```
> act = seq(17, 29, by = 3)
> matplot(act, t(gpa), type = "l", lwd = 2,
+   xlim = c(17, 34))
> legend(30, 3, lty = 1:8, lwd = 2, legend = c("HSR=9", "HSR=8",
+     "HSR=7", "HSR=6", "HSR=5", "HSR=4", "HSR=3", "HSR=2"))
```

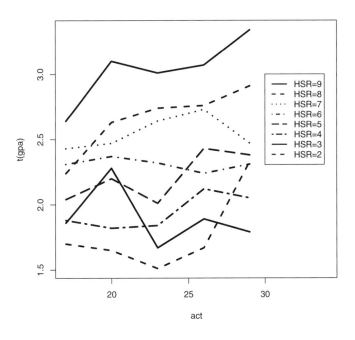

Fig. 10.5. Sample mean GPAs of students for each level of high school rank (HSR) and ACT score.

10.4.2 A Flat Prior Over the Restricted Space

Suppose one is certain before sampling that the population means follow the order restriction, but otherwise one has little opinion about the location of the means. Then if μ denotes the vector of population means, one could assign the flat prior

$$g(\mu) \propto c, \mu \in A,$$

where A is the space of values of μ that follow the order restrictions.

Let y_{ij} and n_{ij} denote the sample mean GPA and sample size, respectively, of the (i, j) cell of the table. We assume that the observations $y_{11}, ..., y_{85}$ are independent with y_{ij} distributed normal with mean μ_{ij} and variance σ^2/n_{ij} where σ is known. The likelihood function of μ then is given by

$$L(\mu) = \prod_{i=1}^{8} \prod_{j=1}^{5} \exp\left\{ -\frac{n_{ij}}{2\sigma^2}(y_{ij} - \mu_{ij})^2 \right\}.$$

Combining the likelihood with the prior, the posterior density is given by

$$g(\mu|y) \propto L(\mu), \mu \in A.$$

This is a relatively complicated 40-dimensional posterior distribution due to the restriction of its mass to the region A. However, to implement the Gibbs sampler, one only requires the availability of the set of full conditional distributions. Here "available" means that the one-dimensional distributions have recognizable distributions that are easy to simulate. Note that the posterior distribution of μ_{ij}, conditional on the remaining components of μ, has the truncated normal form

$$g(\mu_{ij}|y, \{\mu_{jk}, (j,k) \neq (i,j)\}) \propto \exp\left\{ -\frac{n_{ij}}{2\sigma^2}(y_{ij} - \mu_{ij})^2\right\},$$

where $\max\{\mu_{i-1,j}, \mu_{i,j-1}\} \leq \mu_{ij} \leq \min\{\mu_{i,j+1}, \mu_{i+1,j}\}$.

The R function ordergibbs implements Gibbs sampling for this model. As mentioned earlier, we assume that the standard deviation σ is known, and the known value $\sigma = .65$ is assigned inside the function. To begin the algorithm, the program uses a starting value for the matrix of means μ that satisfies the order restriction. Also for ease in programming, the means are embedded within a larger matrix augmented by two rows and two columns containing values of $-\infty$ and $+\infty$. Note in this programming we have changed the ordering of the rows so that the means are increasing from the first to last rows.

```
        [,1]  [,2] [,3] [,4] [,5] [,6] [,7]
 [1,]  -Inf -Inf -Inf -Inf -Inf -Inf -Inf
 [2,]  -Inf 1.59 1.59 1.59 1.67 1.88  Inf
 [3,]  -Inf 1.85 1.85 1.85 1.88 1.88  Inf
 [4,]  -Inf 1.85 1.85 1.85 2.10 2.10  Inf
 [5,]  -Inf 2.04 2.11 2.11 2.33 2.33  Inf
 [6,]  -Inf 2.31 2.33 2.33 2.33 2.33  Inf
 [7,]  -Inf 2.37 2.47 2.64 2.66 2.66  Inf
 [8,]  -Inf 2.37 2.63 2.74 2.76 2.91  Inf
 [9,]  -Inf 2.64 3.02 3.02 3.07 3.34  Inf
[10,]  -Inf  Inf  Inf  Inf  Inf  Inf  Inf
```

In the one main loop, the program goes sequentially through all entries of the population matrix μ, simulating at each step from the posterior of an individual cell mean conditional on the values of the remaining means of the table. The posterior density of μ_{ij} is given by a truncated normal form, where the truncation points depend on the current simulated values of the means in a neighborhood of this (i,j) cell. For example, beginning with the starting value of μ, one would first simulate μ_{11} from a normal $(y_{11}, \sigma/\sqrt{n_{11}})$ distribution truncated on the interval $(-\infty, \min\{1.59, 1.85\})$. As shown in this fragment of the code of the function ordergibbs, a truncated normal simulation is accomplished by the special R function rnormt.

```
lo=max(c(mu[i-1,j],mu[i,j-1]))
hi=min(c(mu[i+1,j],mu[i,j+1]))
mu[i,j]=rnormt(1,y[i-1,j-1],s/sqrt(n[i-1,j-1]),lo,hi)
```

Given the R matrix `iowagpa` containing two columns of sample means and sample sizes, the command `s=ordergibbs(iowagpa,m)` implements Gibbs sampling for m cycles and the matrix of simulated values is stored in the matrix MU. A column of the matrix represents an approximate random sample from the posterior distribution for a single cell mean. In the following, we use $m = 5000$ iterations.

```
> MU = ordergibbs(iowagpa, 5000)
```

The `apply` command is used to find the posterior means of all cell means and the collection of posterior means is placed in an 8-by-5 matrix. Fig. 10.6 displays these posterior means. Note that since the prior support is entirely on the order-restricted space, these posterior means do follow the order restrictions.

```
> postmeans = apply(MU, 2, mean)
> postmeans = matrix(postmeans, nrow = 8, ncol = 5)
> postmeans=postmeans[seq(8,1,-1),]
> dimnames(postmeans)=list(HSR=rlabels,ACTC=clabels)
> round(postmeans,2)
```

	ACTC				
HSR	16-18	19-21	22-24	25-27	28-30
91-99	2.66	2.92	3.01	3.09	3.34
81-90	2.41	2.62	2.73	2.78	2.92
71-80	2.33	2.47	2.62	2.67	2.71
61-70	2.20	2.29	2.33	2.37	2.50
51-60	1.99	2.11	2.15	2.31	2.40
41-50	1.76	1.86	1.94	2.10	2.21
31-40	1.58	1.74	1.80	1.91	2.05
21-30	1.23	1.42	1.55	1.69	1.88

```
> matplot(act, t(postmeans), type = "l", lwd = 2, xlim = c(17, 34))
> legend(30, 3, lty = 1:8, lwd = 2, legend = c("HSR=9", "HSR=8",
+      "HSR=7", "HSR=6", "HSR=5", "HSR=4", "HSR=3", "HSR=2"))
```

One way of investigating the impact of the prior belief in order restriction on inference is to compute the posterior standard deviations of the cell means and compare these estimates with the classical standard errors. By use of the `apply` command, we compute the posterior standard deviations:

```
> postsds = apply(MU, 2, sd)
> postsds = matrix(postsds, nrow = 8, ncol = 5)
> postsds=postsds[seq(8,1,-1),]
> dimnames(postsds)=list(HSR=rlabels,ACTC=clabels)
> round(postsds,3)
```

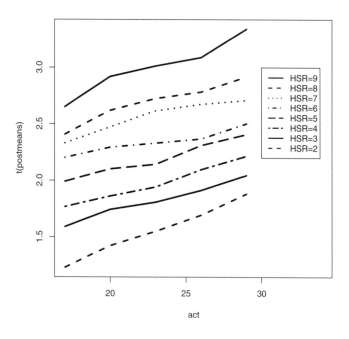

Fig. 10.6. Plot of posterior means of GPAs using noninformative prior on order-restricted space.

	ACTC				
HSR	16-18	19-21	22-24	25-27	28-30
91-99	0.139	0.082	0.053	0.043	0.051
81-90	0.079	0.058	0.038	0.038	0.062
71-80	0.066	0.052	0.038	0.038	0.045
61-70	0.065	0.039	0.035	0.038	0.082
51-60	0.073	0.054	0.055	0.048	0.075
41-50	0.082	0.069	0.068	0.071	0.086
31-40	0.118	0.080	0.074	0.075	0.096
21-30	0.181	0.137	0.118	0.114	0.131

The standard error of the observed sample mean y_{ij} is given by $SE(y_{ij}) = \sigma/\sqrt{n_{ij}}$, where we assume that $\sigma = .65$. The following table computes the ratios $\{SD(\mu_{ij}|y)/SE(y_{ij})\}$ for all cells. Note that most of the ratios are in the .5 to .7 range, indicating that we are adding significant prior information by use of this order-restricted prior.

```
> s=.65
> se=s/sqrt(samplesizes)
> round(postsds/se,2)
```

```
        ACTC
HSR     16-18 19-21 22-24 25-27 28-30
  91-99  0.61  0.49  0.71  0.89  1.00
  81-90  0.54  0.75  0.75  0.78  0.91
  71-80  0.64  0.87  0.79  0.67  0.47
  61-70  0.58  0.58  0.60  0.58  0.55
  51-60  0.72  0.71  0.66  0.57  0.34
  41-50  0.55  0.53  0.63  0.76  0.53
  31-40  0.51  0.37  0.44  0.62  0.44
  21-30  0.56  0.47  0.54  0.58  0.20
```

10.4.3 A Hierarchical Regression Prior

The use of the flat prior over the restricted space A resembles a frequentist analysis where one would find the maximum likelihood estimate. However, from a subjective Bayesian viewpoint, alternative priors could be considered. If one believes that the means satisfy an order restriction, then one may also have prior knowledge about the location of the means. Specifically, one may believe that the mean GPAs may be linearly related to the high school rank and ACT scores of the students.

One can construct a hierarchical regression prior to reflect the relationship between the GPA and the two explanatory variables. At the first stage of the prior, we assume the means are independent where μ_{ij} is normal with location given by the regression structure

$$\beta_0 + \beta_1 \text{ACT}_i + \beta_2 \text{HSR}_j$$

and variance σ_π^2. At the second stage of the prior model, we assume the hyperparameters $\beta = (\beta_0, \beta_1, \beta_2)$ and σ_π^2 are independent with β distributed $N_3(\bar{\beta}, \Sigma_\beta)$ and σ_π^2 distributed $S\chi_\nu^{-2}$.

Prior knowledge about the regression parameter β is expressed by means of the normal prior with mean $\bar{\beta}$ and variance-covariance matrix Σ_β. These values can be obtained by an analysis of similarly classified data for 1978 Iowa students. One can find the MLE and associated variance-covariance matrix from an additive fit to these data. If one assumes that the regression structure between GPA and the covariates has not significantly changed between 1978 and 1990, these values can be used for $\bar{\beta}$ and Σ_β.

To construct a suitable prior for σ_π^2, observe that this parameter reflects the strength of the user's prior belief that the regression model fits the table of observed means. Also this parameter is strongly related to the prior belief that the table of means satisfies the order restriction. The prior mean and standard deviation are given, respectively, by $E(\sigma_\pi^2) = S/(v-2)$ and

$SD(\sigma_\pi^2) = \sqrt{2}/(v-2)/\sqrt{v-4}$. By fixing a value of S and increasing v, the prior for σ_π^2 is placing more of its mass toward zero and reflects a stronger belief in order restriction. In the following, we use the parameter values $S = 0.02$ and $v = 16$.

The posterior density of all parameters $(\mu, \beta, \sigma_\pi^2)$ is given by the following:

$$g(\mu, \beta, \sigma_\pi^2 | y) \propto \prod_{i=1}^{8} \prod_{j=1}^{5} \exp\{-\frac{n_{ij}}{2\sigma^2}(y_{ij} - \mu_{ij})^2\}$$

$$\times \prod_{i=1}^{8} \prod_{j=1}^{5} \frac{1}{\sigma_\pi} \exp\{-\frac{1}{2\sigma_\pi^2}(\mu_{ij} - x'_{ij}\beta)^2\}$$

$$\times \exp\{(\beta - \bar{\beta})'\Sigma^{-1}(\beta - \bar{\beta})\}(\sigma_\pi^2)^{-\nu/2-1} \exp\{-\frac{S}{2\sigma_\pi^2}\}.$$

Simulation from the joint posterior distribution is possible by a Gibbs sampling algorithm. We partition the parameters into the three components μ, β, and σ_π^2, and consider the distribution of each component conditional on the remaining parameters. We describe the set of conditional distributions here; we will see that all of these distributions have convenient functional forms that are easy to simulate on R.

- The population means $\mu_{11}, ..., \mu_{85}$, conditional on β and σ_π^2, are independent $N(\mu_{ij}(y), \sqrt{v_{ij}})$, where

$$\mu_{ij}(y) = v_{ij}\left(\frac{n_{ij}y_{ij}}{\sigma^2} + \frac{x_{ij}\beta}{\sigma_\pi^2}\right), \quad v_{ij} = \left(\frac{n_{ij}}{\sigma^2} + \frac{1}{\sigma_\pi^2}\right)^{-1}.$$

- The regression vector β, conditional on μ and σ_π^2, is distributed $N_3(\beta^*, \Sigma_{\beta^*})$, where

$$\Sigma_{\beta^*} = (\Sigma_\beta^{-1} + X'X\sigma_\pi^{-2})^{-1}, \quad \beta^* = \Sigma_{\beta^*}(\Sigma_\beta^{-1}\bar{\beta} + X'\sigma_\pi^{-2}\mu).$$

- The variance σ_π^2, conditional on μ and β, is distributed according to the inverse gamma form

$$\sigma_\pi^{2^{-(40+v)/2-1}} \exp\{\frac{1}{2\sigma_\pi^2}(S + \sum(\mu_{ij} - x_{ij}\beta)^2)\}.$$

The R function `hiergibbs` implements this Gibbs sampling algorithm. There are two inputs to this function, the data matrix `data` and the number of iterations of the Gibbs sampler `m`. In the program setup, one defines the vector of cell means $\{y_{ij}\}$ (`y`), the vector of sample sizes $\{n_{ij}\}$ (`n`), the design matrix consisting of rows $\{(1, ACT_i, HSR_j)\}$ (`X`) and the vector of known sampling variances $\{\sigma^2/n_{ij}\}$ (`s2`). One defines the prior mean $\bar{b}$ (`b1`), the prior covariance-variance matrix Σ_β (`bvar`), and the hyperparameters of the prior on σ_π^2, S (`s`), and v (`v`). Also, the inverse of Σ_β (`ibar`) is computed.

Before the Gibbs sampling begins, initial values need to be set for the population means $\{\mu_{ij}\}$ and the prior variance σ_π^2. It is convenient to simply let an initial estimate for μ_{ij} be the observed sample mean y_{ij}. Also we let σ_π^2 denote the relatively large value .006 that corresponds to little shrinkage toward the regression model.

We describe the R implementation for a single Gibbs cycle that simulates in turn from the three sets of conditional posterior distributions.

1. **Simulation of β.** This fragment of R code simulates the regression vector β from a multivariate normal distribution. The R command solve is used to compute the inverse of the matrix $\Sigma_\beta^{-1} + X'X\sigma_\pi^{-2}$ and the variance-covariance matrix is stored in the variable pvar. The posterior mean is stored in the variable pmean and the function rmnorm is used to simulate the multivariate normal variate.

```
pvar=solve(ibvar+t(a)%*%a/s2pi)
pmean=pvar%*%(ibvar%*%b1+t(a)%*%mu/s2pi)
beta=rmnorm(1,mean=c(pmean),varcov=pvar)
```

2. **Simulation of σ_π^2.** This R fragment simulates the prior variance from an inverse gamma distribution.

```
s2pi=rigamma(1,(N+v)/2,sum((mu-a%*%beta)^2)/2+s/2)
```

3. **Simulation of μ.** Conditional on the remaining parameters, the components of μ have independent normal distributions. It is convenient to simultaneously simulate all distributions by means of vector operations. The R variable postvar contains values of the posterior variances for the components of μ and postmean contains the respective posterior means. Then the command rnorm(n,postmean,sqrt(postvar)) simulates the values from the 40 independent normal distributions.

```
postvar=1/(1/s2+1/s2pi)
postmean=(y/s2+a%*%beta/s2pi)*postvar
mu=rnorm(n,postmean,sqrt(postvar))
```

The Gibbs sampler is run for 5000 cycles by executing the function hiergibbs.

```
> FIT=hiergibbs(iowagpa,5000)
```

The output variable FIT is a list consisting of three elements: beta, the matrix of simulated regression coefficients β where each row is a simulated draw; mu, the matrix of simulated cell means; and var, the vector of simulated variances σ_π^2.

Fig. 10.7 shows density estimates of the simulated draws of the regression coefficients β_1 and β_2 corresponding respectively to the two covariates high school rank and ACT score. We summarize each coefficient by the computation of the .025, .25, .5, .75, and .975 quantiles of each batch of simulated draws. A 95% interval estimate for β_2, for example, is given by the .025 and .975 quantiles: (.0223, .0346).

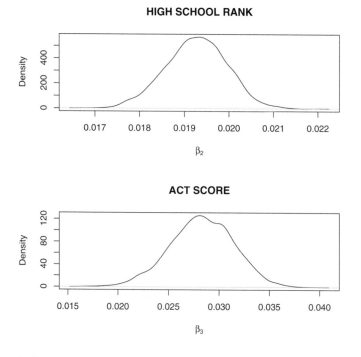

Fig. 10.7. Density estimates of simulated draws of regression coefficients β_1 and β_2 in hierarchical regression model.

```
> par(mfrow=c(2,1))
> plot(density(FIT$beta[,2]),xlab=expression(beta[2]),
+   main="HIGH SCHOOL RANK")
> plot(density(FIT$beta[,3]),xlab=expression(beta[3]),
+   main="ACT SCORE")
> quantile(FIT$beta[,2],c(.025,.25,.5,.75,.975))

       2.5%         25%         50%         75%       97.5%
0.01800818  0.01883586  0.01926438  0.01968747  0.02052101

> quantile(FIT$beta[,3],c(.025,.25,.5,.75,.975))

       2.5%         25%         50%         75%       97.5%
0.02231820  0.02628508  0.02844086  0.03050381  0.03464926
```

We summarize the posterior distribution of the variance parameter σ_π^2; this parameter is helpful for understanding the shrinkage of the observed sample means toward the regression structure.

```
> quantile(FIT$var,c(.025,.25,.5,.75,.975))
```

        2.5%   25%   50%   75%   97.5%
0.001163374 0.002017212 0.002771330 0.003924643 0.007475468

Last, we compute and display the posterior means of the cell means in Fig. 10.8. These posterior mean estimates using a hierarchical prior look similar to the posterior estimates using a noninformative prior on the restricted space displayed in Fig. 10.6.

```
> posterior.means = apply(FIT$mu, 2, mean)
> posterior.means = matrix(posterior.means, nrow = 8, ncol = 5,
+   byrow = T)

> matplot(act, t(posterior.means), type = "l", lwd = 2,
+    xlim = c(17, 34))
> legend(30, 3, lty = 1:8, lwd = 2, legend = c("HSR=9", "HSR=8",
+     "HSR=7", "HSR=6", "HSR=5", "HSR=4", "HSR=3", "HSR=2"))
```

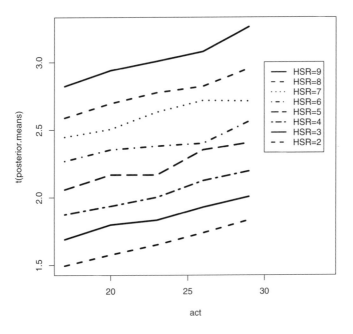

Fig. 10.8. Plot of posterior means of GPAs using hierarchical prior.

10.4.4 Predicting the Success of Future Students

The university is most interested in predicting the success of future students from this model. Let z_{ij}^* denote the college GPA for a single future student with ACT score in class i and high school percentile in class j. If the university believes that a GPA at least 2.5 defines success, then they are interested in computing the posterior predictive probability

$$P(z_{ij}^* \geq 2.5|y).$$

One can express this probability as the integral

$$P(z_{ij}^* \geq 2.5|y) = \int P(z_{ij}^* \geq 2.5|\mu, y)g(\mu|y)d\mu,$$

where $g(\mu|y)$ is the posterior distribution of the vector of cell means μ. In our model, we assume that the distribution of z_{ij}^*, conditional on μ, is $N(\mu_{ij}, \sigma)$. So we can write the predictive probability as

$$P(z_{ij}^* \geq 2.5|y) = \int \left[1 - \Phi\left(\frac{2.5 - \mu_{ij}}{\sigma}\right)\right]g(\mu|y)d\mu,$$

where $\Phi()$ is the standard normal cdf. A simulated sample from the posterior distribution of the cell means is available as one of the outputs of the Gibbs sampling algorithms **ordergibbs** and **hiergibbs**. If $\{\mu_{ij}^t, t = 1, ..., m\}$ represents the sample from the marginal posterior distribution of μ_{ij}, then the posterior predictive probability that the student will be successful is estimated by

$$P(z_{ij}^* \geq 2.5|y) \approx \frac{1}{m}\sum_{t=1}^{m}\left[1 - \Phi\left(\frac{2.5 - \mu_{ij}^t}{\sigma}\right)\right].$$

We illustrate this computation when a hierarchical regression model is placed on the cell means. Recall that the output of the function **hiergibbs** in our example was **FIT** and so **FIT$mu** is the matrix of simulated cell means from the posterior distribution. We transform all the cell means to probabilities of success by use of the **pnorm** function and we compute the sample means for all cells by use of the **apply** function.

```
> p=1-pnorm((2.5-FIT$mu)/.65)
> prob.success=apply(p,2,mean)
```

We convert this vector of estimated probabilities of success to a matrix by the **matrix** command, attach row and column labels to the table by the **dimnames** command, and then display the probabilities, rounding to the third decimal space.

```
> prob.success=matrix(prob.success,nrow=8,ncol=5,byrow=T)
> dimnames(prob.success)=list(HSR=rlabels,ACTC=clabels)
> round(prob.success,3)
```

```
          ACTC
HSR      16-18 19-21 22-24 25-27 28-30
  91-99 0.689 0.748 0.781 0.812 0.878
  81-90 0.555 0.617 0.663 0.690 0.757
  71-80 0.466 0.504 0.579 0.630 0.627
  61-70 0.360 0.410 0.426 0.440 0.538
  51-60 0.249 0.304 0.304 0.410 0.441
  41-50 0.168 0.193 0.222 0.283 0.321
  31-40 0.107 0.141 0.153 0.190 0.225
  21-30 0.062 0.079 0.096 0.121 0.153
```

This table of predictive probabilities should be useful to the admissions officer at the university. By this table, one may wish to admit students who have a predictive probability of, say, at least 0.70, of being successful in college.

10.5 Further Reading

Gelfand and Smith (1990) and Gelfand et al (1990) were the first papers to describe the statistical applications for Gibbs sampling. Wasserman and Verdinelli (1991) and Albert (1992) describe the use of Gibbs sampling in outlier models. The use of latent variables and Gibbs sampling for fitting binary response models is described in Albert and Chib (1993). The use of Gibbs sampling in modeling order restrictions in a two-way table of means was illustrated in Albert (1994).

10.6 Summary of R Functions

bayes.probit – simulates from a probit binary response regression model using data augmentation and Gibbs sampling
Usage: bayes.probit(y, X, m)
Arguments: y, vector of binary responses; X, covariate matrix; m, number of simulations
Value: matrix of simulated draws of the regression vector beta, where each row corresponds to a draw of beta

bprobit.probs – simulates fitted probabilities for a probit regression model
Usage: bprobit.probs(X, fit)
Arguments: X, matrix where each row corresponds to a covariate set; fit, matrix of simulated draws from the posterior distribution of the regression vector beta
Value: matrix of simulated draws of the fitted probabilities, where a column corresponds to a particular covariate set

hiergibbs – implements Gibbs sampling for estimating a two-way table of normal means under a hierarchical regression model

Usage: `hiergibbs(data, m)`
Arguments: `data`, data matrix where columns are observed sample means, sample sizes, and values of two covariates; `m`, number of cycles of Gibbs sampling
Value: `beta`, matrix of simulated values of regression parameter; `mu`, matrix of simulated values of cell means; `var` vector of simulated values of second-stage prior variance

`ordergibbs` – implements Gibbs sampling for estimating a two-way table of normal means under an order restriction
Usage: `ordergibbs(data, m)`
Arguments: `data`, data matrix where first column contains the sample means and the second column contains the sample sizes; `m`, number of iterations of Gibbs sampling
Value: matrix of simulated draws of the normal means where each row represents one simulated draw

`robustt` – implements Gibbs sampling for a robust t sampling model with location mu, scale sigma, and degrees of freedom v
Usage: `robustt(y, v, m)`
Arguments: `y`, vector of data values; `v`, degrees of freedom for t model; `m`, number of cycles of the Gibbs sampler
Value: `mu`, vector of simulated values of mu; `s2`, vector of simulated draws of sigma2; `lam`, matrix of simulated draws of lambda where each row corresponds to a single draw

10.7 Exercises

1. **Robust modeling with Cauchy sampling**
   In Section 6.9, different computational methods are used to model data where outliers may be present. The data $y_1, ...y_n$ are assumed independent, where y_i is Cauchy with location μ and scale σ. Using the standard noninformative prior of the form $g(\mu, \sigma) = 1/\sigma$ and Darwin's dataset, Table 6.2 presents 5th, 50th, and 95th percentiles of the marginal posterior densities of μ and $\log \sigma$ using Laplace, brute force, random walk Metropolis, independence Metropolis, and Metropolis within Gibbs algorithms. Use the "automatic" Gibbs sampler as implemented in the function `robustt` to fit this Cauchy error model where the degrees of freedom of the t density is set to one. Run the algorithm for 10,000 cycles and compute the posterior mean and standard deviation of μ and $\log \sigma$. Compare your answers with the values given in Table 6.2 using the other computational methods.

2. **Mixtures of sampling densities**
   Suppose one observes a random sample $y_1, ..., y_n$ from the mixture density

   $$f(y|p, \lambda_1, \lambda_2) = pf(y|\lambda_1) + (1 - p)f(y|\lambda_2),$$

where $f(y|\lambda)$ is a Poisson density with mean λ, p is a mixture parameter between 0 and 1, and $\lambda_1 < \lambda_2$. Suppose that a priori the parameters $(p, \lambda_1, \lambda_2)$ are independent with p assigned a uniform density and λ_i assigned gamma$(a_i, b_i), i = 1, 2$. Then the joint posterior density is given by

$$g(p, \lambda_1, \lambda_2|\text{data}) \propto g(p, \lambda_1, \lambda_2) \prod_{i=1}^{n} f(y_i|p, \lambda_1, \lambda_2).$$

Suppose one introduces the latent data $Z_1, ..., Z_n$, where $Z_i = 1$ or 2 if $y_i \sim \text{Poisson}(\lambda_1)$ or $y_i \sim \text{Poisson}(\lambda_2)$, respectively. The joint posterior density of the vector of latent data $Z = (Z_1, ..., Z_n)$ and the parameters is given by

$$g(p, \lambda_1, \lambda_2, Z|\text{data}) \propto g(p, \lambda_1, \lambda_2)$$
$$\times \prod_{i=1}^{n} \Big(I(Z_i = 1)pf(y_i|\lambda_1) + I(Z_i = 2)(1-p)f(y_i|\lambda_2)\Big),$$

where $I(A)$ is the indicator function that is equal to 1 if A is true, or 0 otherwise.

a) Find the complete conditional densities of p, λ_1, λ_2, and Z_i.
b) Describe a Gibbs sampling algorithm for simulating from the joint density of $(p, \lambda_1, \lambda_2, Z)$.
c) Write an R function to implement the Gibbs sampler.
d) To test your function, the following data were simulated from the mixture density with $p = .3$, $\lambda_1 = 5$, and $\lambda_2 = 15$:

24 18 21   5   5 11 11 17   6   7
20 13   4 16 19 21   4 22   8 17

Let the prior hyperparameters be equal to $a_1 = b_1 = a_2 = b_2 = 1$. Run the Gibbs sampler for 10,000 iterations. From the simulated output, compute the posterior mean and standard deviation of p, λ_1, and λ_2, and compare the posterior means with the parameter values from which the data were simulated.

3. **Censored data**
Suppose that observations $x_1, ..., x_n$ are normally distributed with mean μ and variance σ^2. However the measuring device has malfunctioned and one only knows the first observation x_1 exceeds a known constant c; the remaining observations $x_2, ..., x_n$ are recorded correctly. If we regard the censored observation x_1 as an unknown and we assign the usual noninformative prior on (μ, σ^2), then the joint density of all unknowns (the single observation and the two parameters) has the form

$$g(\mu, \sigma^2, x_1|\text{data}) \propto \frac{1}{\sigma^2} \prod_{i=2}^{n} \frac{1}{\sqrt{2\pi\sigma^2}} \exp\left\{ -\frac{1}{2\sigma^2}(y_i - \mu)^2\right\}$$
$$\times \frac{1}{\sqrt{2\pi\sigma^2}} \exp\left\{ -\frac{1}{2\sigma^2}(x_1 - \mu)^2\right\}$$

   a) Suppose one partitions the unknowns by $[\mu, \sigma^2]$ and $[x_1]$. Describe the conditional posterior distributions $[\mu, \sigma^2|x_1]$ and $[x_1|\mu, \sigma^2]$.

   b) Write an R function to program the Gibbs sampling algorithm based on the conditional distributions found in part (a).

   c) Suppose the sample observations are 110, 104, 98, 101, 105, 97, 106, 107, 84, 104, where the measuring device is "stuck" at 110 and one knows that the first observation exceeds 110. Use the Gibbs sampling algorithm to find 90% interval estimates for μ and σ,

4. **Order restricted inference**

Suppose one observes $y_1, ..., y_N$, where y_i is distributed binomial with sample size n_i and probability of success p_i. A priori suppose one assigns a uniform prior over the space where the probabilities satisfy the order restriction

$$p_1 < p_2 < ... < p_n.$$

   a) Describe a Gibbs sampling algorithm for simulating from the joint posterior distribution of $(p_1, ..., p_N)$.

   b) Write an R function to implement the Gibbs sampler found in part (a).

   c) Suppose $N = 4$, the sample sizes are $n_1 = ... = n_4 = 20$ and one observes $y_1 = 2$, $y_2 = 5$, $y_3 = 12$, and $y_4 = 9$. Use the R function to simulate 1000 draws from the joint posterior distribution of (p_1, p_2, p_3, p_4).

5. **Grouped data**

In Section 6.7, inference about the mean μ and the variance σ^2 of a normal population is considered, where the heights of male students are observed in grouped form as displayed in Table 6.1. Let $y = (y_1, ..., y_n)$ denote the vector of actual unobserved heights that are distributed $N(\mu, \sigma)$. Consider the joint posterior distribution of all unobservables (y, μ, σ^2). As in Section 6.7, we assume that the parameters (μ, σ^2) have the noninformative prior proportional to $1/\sigma^2$.

   a) Describe the conditional posterior distributions $[y|\mu, \sigma^2]$ and $[\mu, \sigma^2|y]$.

   b) Program an R function that implements a Gibbs sampler based on the conditional posterior distributions found in part (a).

   c) Using the R function, simulate 1000 cycles of the Gibbs sampler. Compute the posterior mean and posterior standard deviation of μ and σ and compare your estimates with the values reported using the Metropolis random walk algorithm in Section 6.7.

11

Using R to Interface with WinBUGS

11.1 Introduction to WinBUGS

The BUGS project is focused on the development of software to facilitate Bayesian fitting of complex statistical models using Markov chain Monte Carlo algorithms. In this chapter, we introduce the use of R in running WinBUGS, a stand-alone software program for the Windows operating system.

WinBUGS is a program for sampling from a general posterior distribution of a Bayesian model by use of Gibbs sampling and a general class of proposal densities. To describe the use of WinBUGS in a very simple setting, suppose you observe y distributed binomial(n, p) and a beta(α, β) prior is placed on p where $\alpha = 0.5$ and $\beta = 0.5$. You observe $y = 7$ successes in a sample of $n = 50$ and you wish to construct a 90% interval estimate for p.

After you launch the WinBUGS program, you create a file that describes the Bayesian model. For this example, the model script looks like the following:

```
model
{
    y ~ dbin(p, n)
    p ~ dbeta( alpha, beta)
}
```

Note that the script begins with **model** and one indicates distributional assumptions by the "~" symbol. The names for different distributions (dbin, dbeta, etc.) are similar to the names of these densities in the R system.

After the model is described, one defines the data and any known parameter values in the file. This script begins with the word **data** and we use a list to specify the values of y, n, α, and β.

```
data
list(y = 7, n = 50, alpha = 0.5, beta = 0.5)
```

Last, we specify the initial values of parameters in the MCMC simulation. This section begins with the word **inits** and a list specifies the initial values.

Here we have a single parameter p and we decide to begin the simulation at $p = .1$.

```
inits
list(p = 0.1)
```

Once the model, data, and initial values have been defined, we tell Win-BUGS, in the Sample Monitor Tool, what parameters to monitor in the simulation. These will be the parameters of primary interest in the inferential problem. Here there is only one parameter p that we wish to monitor.

By use of the Update Tool we are able to use WinBUGS to take a simulated sample of a particular size from the posterior distribution. Once the MCMC simulation is finished, we want to make plots or compute diagnostic statistics of the parameters that help us learn if the MCMC simulation has approximately converged to the posterior distribution. If we believe that the simulation draws represent (approximately) a sample from the posterior, then we want to construct a graph of various marginal posterior distributions of interest and compute various summaries to draw inferences about the parameters.

WinBUGS is useful for fitting a variety of Bayesian models, some of high dimension. But the program runs independently of other programs such as R and one is limited to the data analysis tools available in the WinBUGS system. Recently, there have been efforts to provide interfaces between popular statistical packages (such as R) and WinBUGS. In the remainder of the chapter, we describe one attractive R function `bugs` that simplifies the process of using the WinBUGS program and allows one to use the R system to analyze the simulation output.

11.2 An R Interface to WinBUGS

Before you can use this R/WinBUGS interface, some setup needs to be done. The WinBUGS and OpenBUGS programs should be downloaded and installed on your Windows system. Also, special packages including R2WinBUGS and BRugs need to be downloaded and installed on your R system. This setup procedure likely will be modified over time; you should consult with the Win-BUGS home page (`http://www.mrc-bsu.cam.ac.uk/bugs/`) for the most recent information.

Once the setup is completed, it is easy to define a Bayesian problem for WinBUGS by use of this R interface. There are four necessary inputs that are similar to the inputs required within the WinBUGS program:

- **Model.** One describes the statistical model by means of a "model" file that describes the model in the BUGS language.
- **Data.** One inputs data directly into R in the form of constants, vectors, matrices and model parameters.

- **Parameters.** Within R, one specifies the parameters to be monitored in the simulation run.
- **Initial values.** One specifies initial values of the parameters in the R console.

Suppose the model is defined in the file `model.bug` and the data, parameters, and initial values are defined in R in the respective variables `data`, `parameters` and `inits`. Then one simulates from the Bayesian model by the R command `bugs`:

```
> model.sim <- bugs (data, inits, parameters, "model.bug")
```

When this command is executed, the model information is sent to the WinBUGS program. The WinBUGS program will run in the background, simulating parameters from the model. At the completion of the simulation, WinBUGS will close, and one is returned to the R console. The output of `bugs` is a structure containing the output from the WinBUGS run. Specifically, from the object `model.sim`, one can access the matrix of simulated draws of the monitored parameters.

One controls different aspects of the simulation by use of optional arguments to the function `bugs`. A more general form of `bugs` including optional arguments is given here:

```
bugs(data, inits, parameters.to.save, model.file = "model.bug",
        n.chains = 3, n.iter = 2000, n.burnin = floor(n.iter/2),
        n.thin = max(1, floor(n.chains*(n.iter - n.burnin)/1000)),
        bin = (n.iter - n.burnin) / n.thin)
```

- `n.chains` contains the number of Markov chains that are run. By default, three parallel chains will be run; if one wishes only to simulate one chain, the argument `n.chains = 1` should be used.
- `n.iter` is the number of total iterations for each chain.
- `n.burnin` is the number of iterations to discard at the beginning. Typically, one will discard a specific number of the initial draws and base inference on the remaining output. By default, the first half of the iterations are removed; that is, `n.burnin = n.iter/2`.
- `n.thin` is the thinning rate. If `n.thin = 1`, every iterate will be collected; if `n.thin = 2`, every other iterate will be collected, and so on. By default, the thinning rate is set so that 1000 iterations will be collected for each chain.
- `bin` is the number of iterations between savings of results; the default is only to save at the end.

11.3 MCMC Diagnostics Using the boa Package

Once the MCMC chain has been run and simulated samples from the algorithm have been stored, then the user needs to perform some diagnostics on the simulations to determine if they approximately represent the posterior distribution of interest. Some diagnostic questions include the following:

1. How many chains should be run in the simulation? Does the choice of starting value in the chain make a difference?
2. How long is the burn in time before the simulated draws approximately represent a sample from the posterior distribution?
3. How many simulated draws should be collected to get accurate approximations at summaries of the posterior?
4. What is the simulation standard error of a particular summary of the posterior distribution?
5. Are there high correlations between successive simulated draws?

The boa (Bayesian Output Analysis) package, written by Brian Smith, is based on the Convergence Diagnosis and Output Analysis Software for Gibbs sampling output (CODA) developed by Best, Cowles, and Vines. This package provides a variety of diagnostic functions useful for MCMC output. In particular, the boa package

- provides various summary statistics such as means, standard deviations, quantiles, highest probability density intervals, and simulation standard errors for correlated output based on batch means
- allows one to compare autocorrelations and cross-correlations of simulated samples from different parameters
- computes various convergence diagnostics, such as those proposed by Geweke, Gelman and Rubin, and Raftery and Lewis
- provides a variety of different plots, such as lag correlations, density estimates, and running means

It is convenient to use the boa package after the bugs function is used to perform the MCMC sampling in WinBUGS. One can access the boa functions by the menu option boa.menu(). When the menu appears, one selects Import Data > Data Matrix Object to load vectors or matrices of simulated parameters into the package. Many of the MCMC diagnostics can then be performed by using the Analysis and Plot items on the main menu.

11.4 A Change-Point Model

We begin with an analysis of counts of British coal mining disasters described in Carlin et al (1992). The number of disasters is recorded for each year from 1851 to 1962; we let y_t denote the number of disasters in year t, where $t = $ actual year $- 1850$. Looking at the data, it appears that the rate of accidents decreased in some year during the end of the 19th century. We assume for the early years, say when $t < \tau$, y_t has a Poisson distribution where the logarithm of the mean $\log \mu_t = \beta_0$, and for the later years $(t \geq \tau) \log \mu_t = \beta_0 + \beta_1$. We represent this as

$$y_t \sim \text{Poisson}(\mu_t), \ \log(\mu_i) = \beta_0 + \beta_1 \times \delta(t - \tau),$$

where $\delta()$ is defined to be 1 if its argument is nonnegative, and 0 otherwise. The unknown parameters are the regression parameters β_0, β_1, and the change-point parameter τ. We complete the model by assigning vague uniform priors to β_0 and β_1 and assigning τ a uniform prior on the interval $(1, N)$, where N is the number of years.

The first step in using WinBUGS is to write a short script defining the model in the BUGS language. The description of the change-point model is displayed next. Note that the observation for a particular **year** is denoted by D[year] and the corresponding mean as mu[year]. The parameters are b[1],b[2], and the change-point parameter τ is called **changeyear**. Note that the syntax is similar to that used in R with some exceptions. The syntax

```
D[year] ~ dpois(mu[year])
```

indicates that D[year] is distributed Poisson with mean mu[year]. Similarly, the code

```
b[j] ~ dnorm( 0.0,1.0E-6)
```

indicates that β_j is assigned a normal prior distribution with mean 0 and a precision (reciprocal of the variance) equal to 0.000001. In WinBUGS, one must assign proper distributions to all parameters, and this normal density approximates the improper uniform prior density. Also

```
changeyear ~ dunif(1,N)
```

indicates that τ has a continuous uniform prior density on the interval $(1, N)$. The operator <- indicates an assignment to a variable; for example, the syntax

```
log(mu[year]) <- b[1] + step(year - changeyear) * b[2]
```

assigns the linear expression on the right-hand side to the variable log(mu[year]). The **step** function in WinBUGS is equivalent to the function $\delta()$ defined earlier. The entire model description file is saved as a text file coalmining.bug.

```
model
{
  for( year in 1 : N )
  {
      D[year] ~ dpois(mu[year])
      log(mu[year]) <- b[1] + step(year - changeyear) * b[2]
  }
  for (j in 1:2) {b[j] ~ dnorm( 0.0,1.0E-6)}
  changeyear ~ dunif(1,N)
}
```

After the model has been defined, we enter the data directly into the R console. The R constant N is the number of years and D is the vector of observed counts. The variable data is a list containing the names of the variables N and D that are sent to WinBUGS.

```
> N=112
> D=c(4,5,4,1,0,4,3,4,0,6,
+   3,3,4,0,2,6,3,3,5,4,5,3,1,4,4,1,5,5,3,4,2,5,2,2,3,4,2,1,3,2,
+   1,1,1,1,1,3,0,0,1,0,1,1,0,0,3,1,0,3,2,2,
+   0,1,1,1,0,1,0,1,0,0,0,2,1,0,0,0,1,1,0,2,
+   2,3,1,1,2,1,1,1,1,2,4,2,0,0,0,1,4,0,0,0,
+   1,0,0,0,0,0,1,0,0,1,0,0)
> data=list("N","D")
```

Next we indicate by the parameters line

```
> parameters <- c("changeyear","b")
```

that we wish to monitor the simulated samples of the change-point parameter τ and the regression vector β.

Last, we indicate by the line

```
> inits = function() {list(b=c(0,0),changeyear=50)}
```

that the starting value for the parameter (β_1, β_2) is $(0, 0)$, and the starting value of τ is 50.

Now that the problem has been set up, the function bugs is used to run WinBUGS.

```
> coalmining.sim <- bugs (data, inits, parameters,
+   "coalmining.bug", n.chains=3, n.iter=1000)
```

The output of bugs is the simulation object coalmining.sim.

To obtain some basic information about the simulated draws, one can apply the print function on the simulation object coalmining.sim. The output explains that three chains were used, each with 1000 iterations, and the first 500 iterations (the burn-in) were discarded in each chain. Summary statistics for each parameter are given for the total 1500 iterations that were saved.

```
> print(coalmining.sim)
```

```
Inference for Bugs model at "coalmining.bug", fit using winbugs,
 3 chains, each with 1000 iterations (first 500 discarded)
 n.sims = 1500 iterations saved
```

	mean	sd	2.5%	25%	50%	75%	97.5%	Rhat	n.eff
changeyear	39.5	2.1	36.1	37.8	39.8	40.7	43.6	1	1500
b[1]	1.1	0.1	0.9	1.1	1.1	1.2	1.3	1	350
b[2]	-1.3	0.2	-1.6	-1.4	-1.3	-1.2	-1.0	1	1300
deviance	337.5	2.6	334.2	335.6	336.8	338.6	344.0	1	820

For each parameter, n.eff is a crude measure of effective sample
size, and Rhat is the potential scale reduction factor (at
convergence, Rhat=1).

pD = 3.5 and DIC = 341.0 (using the rule, pD = var(deviance)/2)
DIC is an estimate of expected predictive error (lower deviance
is better).

To be able to work with the simulated samples, one applies the
attach.bugs command on the simulation object. Once this command is applied, the variable changeyear will contain the simulated draws for τ, and b
is a matrix that contains the simulated draws of β_1 and β_2.

```
> attach.bugs(coalmining.sim)
```

We can construct and display a density estimate of the simulated sample
of τ by the plot(density()) command (see Fig. 11.1). This density has an
interesting bimodal shape; this indicates that there is support for a change-
point near 37 and 40 years past 1850.

```
> plot(density(changeyear))
```

Similarly, we can construct density estimates for the simulated draws of
β_1 and β_2. It is clear from Fig. 11.2 that $\beta_2 < 0$ which indicates a drop in the
rate of coal mining facilities beyond the change-point year.

```
> par(mfrow=c(2,1))
> plot(density(b[,1]),xlab="beta1")
> plot(density(b[,2]),xlab="beta2")
```

11.5 A Robust Regression Model

As a second illustration of the R/WinBUGS interface, we consider the fitting
of a robust simple regression model. One is interested in the relationship between the vote count in the 1996 and 2000 presidential elections in the state
of Florida. For each of 67 counties in Florida, one records the voter count for
Pat Buchanan, the Reform party candidate in 2000, and the voter count for
Ross Perot, the Reform party candidate in 1996. Fig. 11.3 plots the square
root of the Buchanan vote against the square root of the Perot count. One
notices a linear relationship with one distinctive outlier. This outlier is due
to an unusual high vote count for Buchanan in Palm Beach County due to a
butterfly ballot design used in that county.

density.default(x = changeyear)

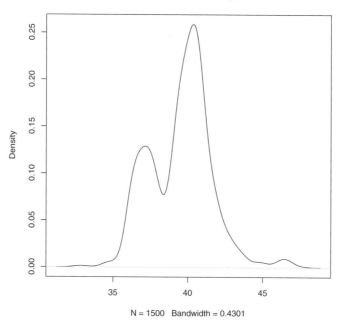

N = 1500   Bandwidth = 0.4301

Fig. 11.1. Density estimate of parameter τ for the change-point problem.

Let y_i and x_i denote the square root of the voter count in the ith county for Buchanan and Perot, respectively. From our preliminary analysis, a linear regression assuming normal errors seems inappropriate. Instead, we assume that $y_1, ..., y_n$ follow the regression model

$$y_i = \beta_0 + \beta_1 x_i + \epsilon_i,$$

where $\epsilon_1, ..., \epsilon_n$ are a random sample from a t distribution with mean 0, scale parameter σ and $\nu = 4$ degrees of freedom. As in Section 10.2, we can represent this model as the following scale mixture of normal distributions:

$$y_i \sim N(\beta_0 + \beta_1 x_i, (\tau \lambda_i)^{-1/2})$$
$$\lambda_i \sim \text{gamma}(2, 2)$$

To complete the model, we assign β_0 and β_1 uniform priors and let the precision τ have the standard noninformative prior proportional to $1/\tau$.

This model is described by means of the following `model` script in Win-BUGS. The observations are `y[1]`, ..., `y[N]`; the observation means are `mu[1]`, ..., `mu[N]`; and the observation precisions are `p[1]`, ..., `p[N]`.

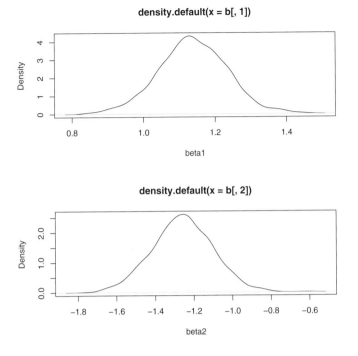

Fig. 11.2. Density estimates of parameters β_1 and β_2 for the change-point problem.

The ith precision, p[i] is defined by tau*lam[i], where the scale parameter lam[i] is assigned a gamma(2, 2) distribution. One cannot formally assign improper priors to parameters, but we approximate a uniform prior for b[1] by assigning it a normal prior with mean 0 and small precision value .001. In a similar fashion, we assign the precision parameter tau a gamma prior with shape and scale parameters each set to the small value of .001. This script is saved as the file robust.bug.

```
model {
for (i in 1:N) {
  y[i] ~ dnorm(mu[i],p[i])
  p[i] <- tau*lam[i]
  lam[i] ~ dgamma(2,2)
  mu[i] <- b[1]+b[2]*x[i]}
for (j in 1:2) {b[j] ~ dnorm(0,0.001)}
tau ~ dgamma(0.001,0.001)
}
```

Next we define the data in R. The Florida voter data for the 1996 and 2000 elections is stored in the dataset election in the package LearnBayes.

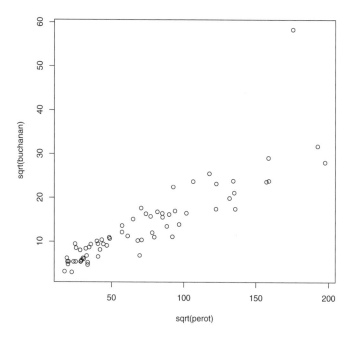

Fig. 11.3. Scatterplot of Buchanan and Perot voter counts in Florida in the 1996 and 2000 presidential elections.

The variables `buchanan` and `perot` contain, respectively, the Buchanan and Perot vote totals. There are three quantities to define, the number of paired observations N, the vector of responses y, and the vector of covariates x. Recall that we applied an initial square root reexpression of both 1996 and 2000 vote totals.

```
> data(election)
> attach(election)
> y=sqrt(buchanan)
> x=sqrt(perot)
> N=length(y)
```

The final two inputs are the selection of initial values for the parameters and the decision on what parameters to monitor in the simulation run. In the command

```
> inits = function() {list(b=c(0,0),tau=1)}
```

we indicate that the starting values for the regression parameters are 0 and 0, and the starting value of the precision parameter τ is 1. We last indicate through the `parameters` statement that we wish to monitor τ, the vector of values $\{\lambda_i\}$, and the regression vector β.

```
> data=list("N","y","x")
> inits = function() {list(b=c(0,0),tau=1)}
> parameters <- c("tau","lam","b")
```

We are ready to use WinBUGS to simulate from the model by the `bugs` function.

```
> robust.sim <- bugs (data, inits, parameters, "robust.bug")
```

Suppose we are interested in estimating the mean Buchanan (root) count $E(y|x)$ for a range of values of the Perot (root) count x. In the R code, we first create a sequence of x values in the variable `xo` and store the corresponding design matrix in the variable `X0`. By multiplying this matrix by the matrix of simulated draws of the regression vector `b`, we get a simulated sample from the posterior of $E(y|x)$ for all values of x in `xo`. We summarize the matrix of posterior distributions `meanresponse` with the 5th, 50th, and 95th percentiles and plot these values as lines in Fig. 11.4. Note that this robust fit is relatively unaffected by the one outlier with an unusually large value of y.

```
> attach.bugs(robust.sim)
> xo=seq(18,196,2)
> X0=cbind(1,xo)
> meanresponse=b%*%t(X0)
> meanp=apply(meanresponse,2,quantile,c(.05,.5,.95))
> lines(xo,meanp[2,])
> lines(xo,meanp[1,],lty=2)
> lines(xo,meanp[3,],lty=2)
```

11.6 Estimating Career Trajectories

A professional athlete's performance level will tend to increase until the middle of his or her career and then deteriorate until retirement. For a baseball player, suppose one records the number of home runs y_j out of the number of balls that are put into play n_j (formally, the number of balls put in play is equal to the number of "at-bats" minus the number of strikeouts) for the jth year of his career. One is interested in the pattern of the home run rate y_j/n_j as a function of the player's age x_j. Fig. 11.5 displays a graph of home run rate against age for the great slugger Mickey Mantle.

To understand a player's career trajectory, we fit a model. Suppose y_j is binomial(n_j, p_j), where p_j is the probability of a home run during the jth season. We assume the probabilities follow the logistic quadratic model

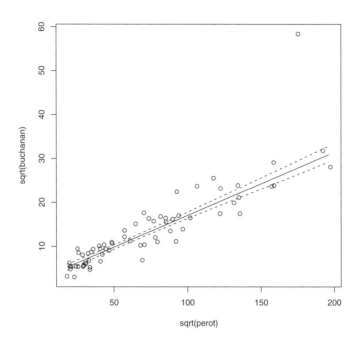

Fig. 11.4. Scatterplot of Buchanan and Perot voter counts. The solid line represents the median of the posterior distribution of the expected response and the dashed lines correspond to the 5th and 95th percentiles of the distribution.

$$\log\left(\frac{p_j}{1-p_j}\right) = \beta_0 + \beta_1 x_j + \beta_2 x_j^2.$$

Fig. 11.5 displays the fitted probabilities for Mickey Mantle using the `glm` function.

In studying a player's career performance, one may be interested in the player's peak ability and the age where he achieved this peak ability. From the quadratic model, if $\beta_2 < 0$, then the probability is maximized at the value

$$age_{PEAK} = -\frac{\beta_1}{2\beta_2}$$

and the peak value of the probability (on the logit scale) is

$$PEAK = \beta_0 - \frac{\beta_1^2}{4\beta_2}.$$

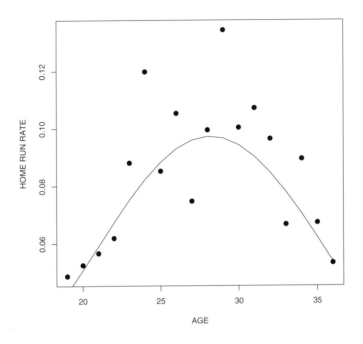

Fig. 11.5. Career trajectory and fitted probabilities for Mickey Mantle's home run rates.

Although fitting this model is informative about a player's career trajectory, it has some limitations. Since a player only plays for 15–20 years and there is sizable binomial variation, it can be difficult to get precise estimates at a player's peak age and his peak ability. But there are many players in baseball history who display similar career trajectories. It would seem that one could obtain improved estimates at players' career trajectories by combining data from players with similar abilities.

One can get improved estimates by fitting an exchangeable model. Suppose we have k similar players; for player i we record the number of home runs y_{ij}, number of balls put in-play n_{ij}, and the age x_{ij} for the seasons $j = 1, ... T_i$. We assume that the associated probabilities $\{p_{ij}\}$ satisfy the logistic model

$$\log\left(\frac{p_{ij}}{1 - p_{ij}}\right) = \beta_{i0} + \beta_{i1}x_{ij} + \beta_{i2}x_{ij}^2, \ j = 1, ..., T_i.$$

Let $\beta_i = (\beta_{i0}, \beta_{i1}, \beta_{i2})$ denote the regression coefficient vector for the ith player. To represent the belief in exchangeability, we assume that $\beta_1, ..., \beta_k$ are a random sample from a common multivariate normal prior with mean vector μ_β and variance-covariance matrix V:

$$\beta_i | \mu_\beta, R \sim N_3(\mu_\beta, V), \quad i = 1, ..., k.$$

At the second stage of the prior, we assign vague priors to the hyperparameters.

$$\mu_\beta \sim c, \quad V \sim \text{inverse Wishart}(S^{-1}, \nu),$$

where inverse Wishart(S^{-1}, ν) denotes the inverse Wishart distribution with scale matrix S and degrees of freedom ν. In WinBUGS, information about a variance-covariance matrix is represented by means of a Wishart(S, ν) distribution placed on the precision matrix P:

$$P = V^{-1} \sim \text{Wishart}(S, \nu).$$

Data are available for 10 great home run hitters in baseball history in the dataset `sluggerdata` in the package LearnBayes. This dataset contains batting statistics for these players for all seasons of their careers. The R function `careertraj.setup` is used to extract the matrices from `sluggerdata` that will be used in the WinBUGS program.

```
> data(sluggerdata)
> s=careertraj.setup(sluggerdata)
> N=s$N; T=s$T; y=s$y; n=s$n; x=s$x
```

The variable N is the number of players and the vector T contains the number of seasons for each player. The matrix y has 10 rows and 23 columns where the ith row in y represents the number of home runs of the ith player for the years of his career. Similarly, the matrix n contains the number of balls put in play for all players and the matrix x contains the ages of the players for all seasons.

A listing of the file `career.bug` describing the model in the WinBUGS language is shown next. The variable beta is a matrix where the ith row corresponds to the regression vector for the ith player. The syntax

```
beta[i , 1:3] ~ dmnorm(mu.beta[], R[ , ])
```

indicates that the i row of beta is assigned a multivariate normal prior with mean vector mu.beta and precision matrix R. The syntax

```
y[i,j] ~ dbin(p[i,j],n[i,j])
logit(p[i,j])<-beta[i,1]+beta[i,2]*x[i,j]+
              beta[i,3]*x[i, j]*x[i, j]
```

gives the logistic model for the home run probabilities in the matrix p. Finally, the syntax

```
mu.beta[1:3] ~ dmnorm(mean[1:3],prec[1:3 ,1:3 ])
R[1:3 , 1:3] ~ dwish(Omega[1:3 ,1:3 ], 3)
```

assigns the second-stage priors. The mean vector mu.beta is assigned a multivariate normal prior with mean mean and precision matrix prec; the precision matrix R is assigned a Wishart distribution with scale matrix Omega and degrees of freedom 3.

```
model
{
for( i in 1 : N ) {
  b
  for( j in 1 : T[i] ) {
    y[i,j] ~ dbin(p[i,j],n[i,j])
    logit(p[i,j])<-beta[i,1]+beta[i,2]*x[i,j]+
                  beta[i,3]*x[i, j]*x[i, j]
  }
}
mu.beta[1:3] ~ dmnorm(mean[1:3],prec[1:3 ,1:3 ])
R[1:3 , 1:3] ~ dwish(Omega[1:3 ,1:3 ], 3)
}
```

The dataset variables N, T, y, n, and x have already been defined in R with help of the `careertraj.setup` function. One defines the hyperparameter values at the last stage of the prior.

```
mean = c(0, 0, 0)
Omega=diag(c(.1,.1,.1))
prec=diag(c(1.0E-6,1.0E-6,1.0E-6))
```

Next one gives initial estimates for β, μ_β, and R. The estimate of β_i is found by fitting a logistic model to the pooled dataset for all players and μ_β is also set to be this value. The precision matrix R is initially given a diagonal form with small values.

```
beta0=matrix(c(-7.69,.350,-.0058),nrow=10,ncol=3,byrow=TRUE)
mu.beta0=c(-7.69,.350,-.0058)
R0=diag(c(.1,.1,.1))
```

We then indicate in the **data** line the list of variables, the **inits** function specifies the initial values and the **parameter** line indicates that we will monitor only the matrix **beta**. We run the MCMC simulation by the **bugs** command.

```
data=list("N","T","y","n","x","mean","Omega","prec")
inits = function() {list(beta=beta0,mu.beta=mu.beta0,R=R0)}
parameters <- c("beta")
career.sim <- bugs (data, inits, parameters, "career.bug",
  n.chains=1, n.iter=10000, n.thin=1)
```

Since we saved the output in the variable **career.sim**, the simulated draws of β are contained in the component **career.sims$sims.list$beta**. This is a three-dimensional array, where **beta[,i,1]** contains the simulated draws of β_{i0}, **beta[,i,2]** contains the simulated draws of β_{i1}, and **beta[,i,3]** contains the simulated draws of β_{i2}. Suppose we focus on the estimates of the peak age for each player. In the following R code, we create a new matrix to hold the simulated draws of the peak age and then compute the functions in a loop.

```
peak.age=matrix(0,5000,10)
for (i in 1:10)
  peak.age[,i]=-career.sim$sims.list$beta[,i,2]/2/
    career.sim$sims.list$beta[,i,3]
```

We illustrate the use of the `boa` package to perform output analysis and summarize the samples of peak age parameters. We invoke the `boa` menu system by typing

```
boa.menu()
```

```
BOA MAIN MENU
*************

1: File      >>
2: Data      >>
3: Analysis >>
4: Plot      >>
5: Options  >>
6: Window    >>
```

To read in the matrix `peak.age` into the package, we choose the menu option `File -> Import Data -> Data Matrix Object` and entered the object name `peak.age`. To obtain trace plots for each parameter, we return to the main menu and choose the menu option `Plot -> Descriptive -> Trace`. Fig. 11.6 displays the trace plots that are produced for the peak age parameters for the first six players. In a similar fashion, one can produce alternative graphs such as autocorrelations or running means. Density plots for the parameters can be obtained by the menu selection `Plot -> Descriptive -> Density`. Fig. 11.7 displays density estimates of the peak ages for the same six players. Finally, to compute 95% interval estimates of each parameter, we use the menu option `Analysis -> Descriptive Statistics -> Highest Probability Density Intervals` and the following output is produced:

	Lower Bound	Upper Bound
par1	31.03141	35.65860
par10	28.35016	30.92325
par2	29.68170	34.69904
par3	31.17224	33.74842
par4	26.49693	29.03557
par5	29.27239	31.55626
par6	26.90644	29.53686
par7	27.74544	30.79067
par8	29.99571	32.96604
par9	26.53799	28.93148

We see that baseball players generally peak in home run hitting ability in their early 30s, although there are some exceptions.

Sampler Trace

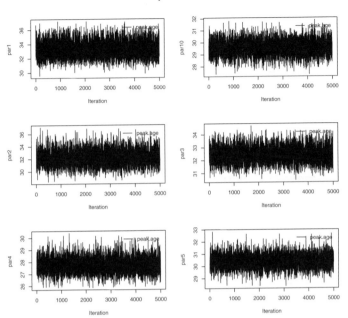

Fig. 11.6. Trace plots of the peak age parameters for six of the baseball players.

11.7 Further Reading

Cowles (2004) gives a general review and evaluation of WinBUGS. A tutorial on computing Bayesian analyses via WinBUGS is provided by George Woodworth in the complement of chapter 6 of Press (2003). General information about WinBUGS including the program code for many examples can be found in the WinBUGS user manual Spiegelhalter et al (2003). Congdon (2003, 2005, 2007) describes a wide variety of Bayesian inference problems that can be fit using WinBUGS. Cowles and Carlin (1996) give an overview of diagnostics for MCMC output. Sturtz et al (2005) give a general description of the R2WinBUGS package including examples demonstrating the use of the package. Smith (2004) describes the use of the package BOA in implementing MCMC output analysis on R.

Estimated Posterior Density

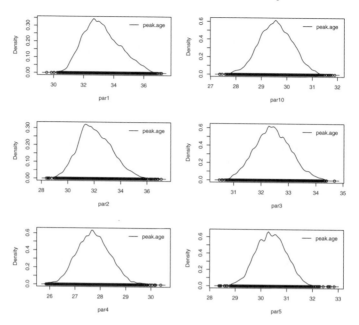

Fig. 11.7. Density estimates of the peak age parameters for six of the baseball players.

11.8 Exercises

1. **Estimation of a proportion with a discrete prior**

   In Chapter 2, we considered the situation where one observes $y \sim$ binomial(n, p) and the proportion p is assigned a discrete prior. Suppose the possible values of p are .05, .15, ..., .95, with respective prior probabilities .0625, .125, .25, .25, .125, .0625, .03125, .03125, .03125, .03125. Place the values of p in a vector `p` and the probabilities in the vector `prior`. As in the example of Chapter 2, set $y = 11$ and $n = 27$. Define `data`, `inits`, and `parameters` as follows:

   ```
   data=list("p","prior","n","y")
   inits=function() {list(ind=2)}
   parameters=list("prob")
   ```

   Save the following script in a file "proportion.bug".

   ```
   model
   {
   ind~dcat(prior[])
   ```

```
prob<-p[ind]
y~dbin(prob,n)
}
```

Use the R interface to simulate 1000 draws from the posterior distribution of p. Compute the posterior probability that p is larger than .5.

2. **Fitting an binomial/beta exchangeable model**

   In Chapter 5, we considered the problem of simultaneously estimating the rates of death from stomach cancer for males at risk for cities in Missouri. Assume the number of cancer deaths y_j for a given city is binomial with sample size n_j and probability of success p_j. To model the belief that the $\{p_j\}$ are exchangeable, we assume that they are a random sample from a beta(α, β) distribution. The beta parameters α and β are assumed independent from gamma(.11,.11) distributions. The WinBUGS model file is shown here. Note that the variable betamean is the prior mean of p_j and K1 is the prior precision.

```
model
{
for (i in 1:N) {
        y[i] ~ dbin( p[i], n[i] )
        p[i] ~ dbeta( alpha, beta )
}
alpha ~ dgamma(.11, .11)
beta ~ dgamma(.11, .11)
betamean <- alpha /( alpha + beta)
K1<-alpha+beta;
}
```

   Use the R interface to simulate from the joint posterior distribution of $(\{p_j\}, \alpha, \beta)$. Summarize each probability p_j and the prior mean $\alpha/(\alpha+\beta)$ and prior precision $K = \alpha + \beta$ by 90% interval estimates.

3. **Smoothing multinomial counts**

   Consider the observed multinomial frequencies (14, 20, 20, 13, 14, 10, 18, 15, 11, 16, 16, 24). Using a GLIM formulation for these data, suppose that the counts $\{y_i\}$ are independent Poisson with means $\{\mu_i\}$. The multinomial proportion parameters are defined by $\theta_i = \mu_i/\sum_j \mu_j$. Suppose one believes that the $\{\theta_i\}$ are similar in size. To model this belief, assume that $\{\theta_i\}$ has a symmetric Dirichlet distribution of the form

$$g(\{\theta_i\}|k) \propto \prod_{i=1}^{12} \theta_i^{k-1}.$$

   The hyperparameter k has a prior density proportional to $(1 + k)^{-2}$ that is equivalent to $\log k$ distributed according to a standard logistic distribution. The WinBUGS model description is shown here:

```
model
{
logk~dlogis(0,1)
k<-exp(logk)
for (i in 1:I) { mu[i]  ~ dgamma(k,1)
                 x[i]  ~ dpois(mu[i])
                 theta[i] <- mu[i]/mu.sum }
mu.sum <- sum(mu[]);
}
```

Using the R interface, simulate from the posterior distribution of $\{\theta_i\}$ and K. Summarize each parameter by a posterior mean and standard deviation.

4. **A gamma regression model**

Congdon (2007) gives a Bayesian analysis of an example from McCullagh and Nelder (1989) modeling the effects of three nutrients on coastal Bermuda grass. The design was a $4 \times 4 \times 4$ factorial experiment defined by replications involving the nutrients nitrogen (N), phosphorus (P), and potassium (K). The response y_i is the yield of grass in tons/acre. We assume y_i is gamma with shape ν and scale parameter $\nu \epsilon_i$ where the mean ϵ_i satisfies

$$1/\epsilon_i = \beta_0 + \beta_1/(N_i + \alpha_1) + \beta_2/(P_i + \alpha_2) + \beta_3/(K_i + \alpha_3).$$

In Congdon's formulation, α_1, α_2, and α_3 (background nutrient levels) are assigned independent normal priors with respective means 40, 22, and 32 and variance 100. Noninformative priors were assigned to β_0, ν and the growth effect parameters β_1, β_2, and β_3, except that the growth effects are assumed to be positive.

The WinBUGS model description is shown here. The LearnBayes data file `bermuda.grass` contains the data; the factor levels are stored in the variables `Nit`, `Phos`, and `Pot` and the response values are stored in the variable y. Also one needs to define the sample size variable n = 64 and the nutrient value vectors N= 0, 100, 200, and 400, P= 0, 22, 44, and 88, and K =0, 42, 84, and 168.

```
model    {for (i in 1:n) {y[i]~ dgamma(nu,mu[i])
        mu[i] <- nu*eta[i]
        yhat[i] <- 1/eta[i]
        eta[i] <- beta0
                  +beta[1]/(N[Nit[i]+1]+alpha[1])
                  +beta[2]/(P[Phos[i]+1]+alpha[2])
                  +beta[3]/(K[Pot[i]+1]+alpha[3])}
beta0 ~ dnorm(0,0.0001)
nu ~ dgamma(0.01,0.01)
alpha[1] ~ dnorm(40,0.01)
```

```
alpha[2] ~ dnorm(22,0.01)
alpha[3] ~ dnorm(32,0.01)
for (j in 1:3) {beta[j] ~ dnorm(0,0.0001) I(0,)}}
```

Use WinBUGS and the R interface to simulate 10,000 iterations from this model. Compute 90% interval estimates for all parameters.

5. **A non-linear hierarchical growth curve model**

The BUGS manual presents an analysis of data originally presented in Draper and Smith (1998). The response y_{ij} is the trunk circumference recorded at time $x_j = 1, ..., 7$ for each of $i = 1, ..., 5$ orange trees; the data are displayed in Table 11.1. One assumes y_{ij} is normally distributed with mean η_{ij} and variance σ^2 where the means satisfy the nonlinear growth model

$$\eta_{ij} = \frac{\phi_{1i}}{1 + \phi_{i2}\exp(\phi_{i3}x_j)}.$$

Suppose one reexpresses the parameters into the real-valued parameters

$$\theta_{i1} = \log\phi_{i1}, \ \theta_{i2} = \log(\phi_{i2}+1), \ \theta_{i3} = \log(-\phi_{i3}), \ i = 1, ...5.$$

Table 11.1. Data on the growth of five orange trees over time.

	Response for Tree Number				
x	1	2	3	4	5
118	30	33	30	32	30
484	58	69	51	62	49
664	87	111	75	112	81
1004	115	156	108	167	125
1231	120	172	115	179	142
1372	142	203	138	209	174
1582	145	203	140	214	177

Let $\theta_i = (\theta_{i1}, \theta_{i2}, \theta_{i3})$ represent the vector of growth parameters for the ith tree. To reflect a prior belief in similarity in the growth patterns of the five trees, one assumes that $\{\theta_i, i = 1, ..., 5\}$ are a random sample from a multivariate normal distribution with mean vector μ and variance-covariance matrix Ω. At the final stage of the prior, one assumes Ω^{-1} is Wishart with parameters R and 3, and assumes μ is multivariate normal with mean vector μ_0 and variance-covariance matrix M. In this example, one assumes R is a diagonal matrix with diagonal elements .1, .1, and .1, μ_0 is the zero vector, and M^{-1} is the diagonal matrix with diagonal elements 1.0E-.6, 1.0E-6, and 1.0E-6. The WinBUGS model description is shown here:

```
model {
    for (i in 1:K) {
    for (j in 1:n) {
        Y[i, j] ~ dnorm(eta[i, j], tauC)
        eta[i, j] <- phi[i, 1] / (1 + phi[i, 2] *
         exp(phi[i, 3] * x[j]))
            }
        phi[i, 1] <- exp(theta[i, 1])
        phi[i, 2] <- exp(theta[i, 2]) - 1
        phi[i, 3] <- -exp(theta[i, 3])
        theta[i, 1:3] ~ dmnorm(mu[1:3], tau[1:3, 1:3])
    }
    mu[1:3] ~ dmnorm(mean[1:3], prec[1:3, 1:3])
    tau[1:3, 1:3] ~ dwish(R[1:3, 1:3], 3)
    sigma2[1:3, 1:3] <- inverse(tau[1:3, 1:3])
    for (i in 1 : 3) {sigma[i] <- sqrt(sigma2[i, i]) }
    tauC ~ dgamma(1.0E-3, 1.0E-3)
    sigmaC <- 1 / sqrt(tauC)
    }
```

Use WinBUGS and the R interface to simulate 10,000 iterations from this model. Compute 90% interval estimates for all parameters.

References

Agresti, A., and Franklin, C. (2005), *Statistics: The Art and Science of Learning from Data*, Prentice-Hall.

Albert, J. (1992), "A Bayesian analysis of a Poisson random effects model for home run hitters," *The American Statistician*, 46, 246-253.

Albert, J. (1994), "A Bayesian approach to estimation of GPA's of University of Iowa freshmen under order restrictions," *Journal of Educational Statistics*, 19, 1-22.

Albert, J. (1996), *Bayesian Computation Using Minitab*, Belmont, CA: Duxbury Press.

Albert, J., and Chib, S. (1993), "Bayesian analysis of binary and polychotomous response data," *Journal of the American Statistical Association*, 88, 669-679.

Albert, J., and Gupta, A. (1981), "Mixtures of Dirichlet distributions and estimation in contingency tables," *Annals of Statistics*, 10, 1261-1268.

Albert, J., and Rossman, A. (2001), *Workshop Statistics: Discovery with Data, a Bayesian Approach*, Emeryville, CA: Key College.

Antleman, G. (1996), *Elementary Bayesian Statistics*, Cheltenham: Edward Elgar Publishing.

Berger, J. (1985), *Statistical Decision Theory and Bayesian Analysis*, New York: Springer-Verlag.

Berger, J. (2000), "Bayesian analysis: A look at today and thoughts of tomorrow," *Journal of the American Statistical Association*, 95, 1269-1276.

Berger, J., and Sellke, T. (1987), "Testing a point null hypothesis: The irreconcilability of p values and evidence," *Journal of the American Statistical Association*, 397, 112-122.

Berry, D. (1996), *Statistics: A Bayesian Perspective*, Belmont, CA: Duxbury Press.

Bliss, C. (1935), "The calculation of the dosage-mortality curve," *Annals of Applied Biology*, 22, 134-167.

Bolstad, W. (2004), *Introduction to Bayesian Statistics*, Hoboken, NJ: John Wiley.

Box, G. (1980), "Sampling and Bayes' inference in scientific modelling and robustness (with discussion)," *Journal of the Royal Statistical Society, Series A*, 143, 383-430.

Carlin, B., Gelfand, A. and Smith, A. (1992), "Hierarchical Bayesian analysis of changepoint problems," *Applied Statistics*, 41, 389-405.

Carlin, B., and Louis, T. (2000), *Bayes and Empirical Bayes Methods for Data Analysis*, Boca Rotan: Chapman and Hall.

Casella, G., and Berger, R. (1987), "Testing a point null hypothesis: The irreconcilability of p values and evidence," *Journal of the American Statistical Association*, 397, 106-111.

Casella, G., and George, E. (1992), "Explaining the Gibbs sampler," *The American Statistician*, 46, 167-174.

Chaloner, K., and Brant, R. (1988), "A Bayesian approach to outlier detection and residual analysis," *Biometrika*, 75, 651-659.

Chib, S., and Greenberg, E. (1995), "Understanding the Metropolis-Hastings algorithm," *The American Statistician*, 49, 327-335.

Christiansen, C., and Morris, C. (1995), "Fitting and checking a two-level Poisson model: modeling patient mortality rates in heart transplant patients," in Berry, D. and Stangl, D, eds, *Bayesian Biostatistics*, New York: Marcel Dekker.

Collett, D. (1994), *Modelling Survival Data in Medical Research*, London: Chapman and Hall.

Congdon, P. (2007), *Bayesian Statistical Modelling*, second edition, Chichester: John Wiley.

Congdon, P. (2004), *Applied Bayesian Modelling*, Chichester: John Wiley.

Congdon, P. (2005), *Bayesian Models for Categorical Data*, Chichester: John Wiley.

Cowles, K. (2004), "Review of WinBUGS 1.4," *The American Statistician*, 58, 330–336.

Cowles, K., and Carlin, B. (1996), "Markov chain Monte Carlo convergence diagnostics: a comparative review," *Journal of the American Statistical Association*, 91, 883–904.

Dobson, A. (2001), *An Introduction to Generalized Linear Models*, New York: Chapman and Hall.

Draper, N., and Smith, H. (1998), *Applied Regression Analysis*, New York: John Wiley.

Edmonson, J., Fleming, T., Decker, D., Malkasian, G., Jorgensen, E., Jefferies, J., Webb, M, and Kvols, L. (1979), "Different chemotherapeutic sensitivities and host factors affecting prognosis in advanced ovarian carcinoma versus minimal residual disease," *Cancer Treatment Reports*, 63, 241–247.

Fisher, R. (1960), *Statistical Methods for Research Workers*, Edinburgh: Oliver & Boyd.

Gelfand, A., and Smith, A. (1990), "Sampling-based approaches to calculating marginal densities," *Journal of the American Statistical Association*, 85, 398–409.

Gelfand, A., Hills, S., Racine-Poon, A., and Smith, A. (1990), "Illustration of Bayesian inference in normal data models using Gibbs sampling," *Journal of the American Statistical Association*, 85, 972–985.

Gelman, A., Carlin, J., Stern, H. and Rubin, D. (2003), *Bayesian Data Analysis*, New York: Chapman and Hall.

Gelman, A., Meng, X. and Stern, H. (1996), "Posterior predictive assessment of model fitness via realized discrepancies," *Statistics Sinica*, 6, 733–807.

Gentle, J. (2002), *Elements of Computational Statistics*, New York: Springer.

Gilchrist, W. (1984), *Statistical Modeling,* Chichester: John Wiley and Sons.

Gill, J. (2002), *Bayesian Methods,*, New York: Chapman and Hall.

Givens, G., and Hoeting, J. (2005), *Computational Statistics*, Hoboken, NJ: John Wiley.

Grayson, D. (1990), "Donner party deaths: a demographic assessment," *Journal of Anthropological Assessment*, 46, 223–242.

Gunel, E., and Dickey, J. M. (1974), "Bayes factors for independence in contingency tables," *Biometrika*, 61, 545–557.

Haberman, S. (1978), *Analysis of Qualitative Data: Introductory topics (Vol. 1)*, New York: Academic Press.

Hartley, H. O. (1958), "Maximum likelihood estimation from incomplete data," *Biometrics*, 14, 174–194.

Howard, J. (1998), "The 2 × 2 table: a discussion from a Bayesian viewpoint," *Statistical Science*, 13, 351–367.

Kass, R., and Raftery, A. (1995), "Bayes factors," *Journal of the American Statistical Association*, 90, 773–795.

Kemeny, J., and Snell, J. (1976), *Finite Markov Chains*," New York: Springer-Verlag.

Lee, P. (2004), *Bayesian Statistics: An Introduction*, New York: Oxford University Press.

Martz, H., and Waller, R. (1982), *Bayesian Reliability Analysis*, New York: John Wiley.

McCullagh, P., and Nelder, J. (1989), *Generalized Linear Models*, New York: Chapman and Hall.

Monahan, J. (2001), *Numerical Methods of Statistics*, Cambridge: Cambridge University Press.

Moore, D. (1995), *The Basic Practice of Statistics*, New York: W. H. Freeman.

Pearson, E. (1947), "The choice of statistical tests illustrated in the interpretation of data classed in a 2 x 2 table," *Biometrika*, 34, 139–167.

Pimm, S., Jones, H., and Diamond, J. (1988), "On the risk of extinction," *American Naturalist*, 132, 757–785.

Press, J. (2003), *Subjective and Objective Bayesian Statistics,* Hoboken, NJ: John Wiley.

Ramsey F., and Schafer, D. (1997), *The Statistical Sleuth*, Belmont CA: Duxbury Press.

Rao, C. R. (2002), *Linear Statistical Inference and Applications*, New York: John Wiley & Sons.

Robertson, T., Wright, F. and Dykstra, R. (1988), *Order Restricted Statistical Inference,* London: John Wiley.

Robert, C., and Casella, G. (2004), *Monte Carlo Statistical Methods*, New York: Springer.

Smith, B. (2004), "boa: Bayesian output analysis program (BOA) for MCMC," R package version 1.1.2-1, URL `http://www.public-health.uiowa.edu/boa`.

Smith, A., and Gelfand, A. (1992), "Bayesian statistics without tears: a sampling-resampling perspective," *The American Statistician*, 46, 84–88.

Spiegelhalter, D., Thomas, A., Best, N., and Lunn, D. (2003), *WinBUGS 1.4 Manual*.

Sturtz, S., Ligges, U., and Gelman, A. (2005), "R2WinBUGS: A package for running WinBUGS from R," *Journal of Statistical Software*, 12, 1–16.

Tanner, M. (1996), *Tools for Statistical Inference*, New York: Springer-Verlag.

Tsutakawa, R., Shoop, G., and Marienfeld, C. (1985), "Empirical Bayes Estimation of Cancer Mortality Rates," *Statistics in Medicine*, 4, 201–212.

Turnbull, B., Brown, B. and Hu, M. (1974), "Survivorship analysis of heart transplant data," *Journal of the American Statistical Association*, 69, 74–80.

Verzani, J. (2004), *Using R for Introductory Statistics*, Boca Raton: Chapman and Hall.

Wasserman, L., and Verdinelli, I. (1991), "Bayesian analysis of outlier models using the Gibbs sampler," *Statistics and Computing*, 1, 105–117.

Weiss, N. (2001), *Elementary Statistics*, Boston: Addison-Wesley.

Index

Bayesian Core: A Practical Approach to Computational Bayesian Statistics

Jean-Michel Marin and Christian P. Robert

This Bayesian modeling book is intended for practitioners and applied statisticians looking for a self-contained entry to computational Bayesian statistics. Focusing on standard statistical models and backed up by discussed real datasets available from the book website, it provides an operational methodology for conducting Bayesian inference, rather than focusing on its theoretical justifications. Special attention is paid to the derivation of prior distributions in each case and specific reference solutions are given for each of the models.

2007. 270 pp. (Springer Texts in Statistics) Hardcover
ISBN 978-0-387-38979-0

Pattern Recognition and Machine Learning

Christopher M. Bishop

The dramatic growth in practical applications for machine learning over the last ten years has been accompanied by many important developments in the underlying algorithms and techniques. This completely new textbook reflects these recent developments while providing a comprehensive introduction to the fields of pattern recognition and machine learning. It is aimed at advanced undergraduates or first-year PhD students, as well as researchers and practitioners. No previous knowledge of pattern recognition or machine learning concepts is assumed.

2006. 702 pp. (Information Science and Statistics) Hardcover
ISBN 978-0-387-31073-2

Model-based Geostatistics

Peter J. Diggle and Paulo Justiniano Ribeiro

Model-based geostatistics refers to the application of general statistical principles of modeling and inference to geostatistical problems. This volume is the first book-length treatment of model-based geostatistics. The book assumes a working knowledge of classical and Bayesian methods of inference, linear models, and generalized linear models, but does not require previous exposure to spatial statistical models or methods. The authors have used the material in MSc-level statistics courses.

2006. 230 pp. (Springer Series in Statistics) Hardcover
ISBN 978-0-387-32907-9

Printed in the United States of America